DARKNESS
DESIGN AND BIODIVERSITY

Eliminating Light Pollution in Architecture: A Multidisciplinary and Integrated Approach Towards the Suitable Design of Nocturnal Spaces

JULLE OKSANEN, PhD

Copyright © 2024 Julle Oksanen

Cataloguing data available from Library and Archives Canada
978-0-88839-786-7 [paperback]
978-0-88839-787-4 [epub]

All rights reserved. No part of this publication may be reproduced, stored in a retrieval system or transmitted, in any form or by any means, electronic, mechanical, audio, photo-copying, recording, or otherwise (except for copying permitted by Sections 107 and 108 of the U.S. Copyright Law and except for book reviews for the public press), without the prior written permission of Hancock House Publishers. Permissions and licensing contribute to the book industry by helping to support writers and publishers through the purchase of authorized editions and excerpts.
Please visit www.accesscopyright.ca.

Photographs are copyrighted and taken by the author unless otherwise stated.

Printed in South Korea

FRONT COVER DESIGN J. RADE
PRODUCTION & DESIGN by J. Rade, M. Lamont
EDITING BY D. MARTENS

We acknowledge the support of the Government of Canada through the Canada Book Fund and the Canada Council for the Arts, and of the Province of British Columbia through the British Columbia Arts Council and the Book Publishing Tax Credit.

Hancock House gratefully acknowledges the Halkomelem Speaking Peoples whose unceded, shared and asserted traditional territories our offices reside upon.

Published simultaneously in Canada and the United States by
HANCOCK HOUSE PUBLISHERS LTD.
19313 Zero Avenue, Surrey, B.C. Canada V3Z 9R9
#104-4550 Birch Bay-Lynden Rd, Blaine, WA, U.S.A. 98230-9436
(800) 938-1114 Fax (800) 983-2262
www.hancockhouse.com info@hancockhouse.com

*For my family, wife Leila and our children
Olli, Pauliina, Karoliina and Petteri.*

Thank you for your continuous and unlimited support.

ACKNOWLEDGEMENTS

Antti Ahlava

Vociferous thanks to Dr. Antti Ahlava, professor of developing design methods at Aalto University's Department of Architecture, for his all-embracing supervision, not only of my doctoral studies but also of my postdoctoral endeavors, and for his part in bringing this book to fruition.

Hannu Tikka

Equally vociferous thanks are due to Professor and Doctor of Architecture Hannu Tikka for over thirty years of cooperation, with whom we joined forces in forging an international career in the promotion of architecture and lighting. Over more than ten years we have proclaimed light and space worldwide at more than twenty universities, where we have a total of five professorships. Without Hannu, I would never have achieved a doctorate. The road we have shared has also made us kindred spirits and good friends.

Daniel Klem, Jr.

Warm and cordial thanks to the internationally renowned Professor Daniel Klem, Jr., who in the process of writing this book became for me an important fatherly mentor, an eminent source of expert help and a great moral support. Professor Klem is the Sarkis Acopian Professor of Ornithology and Conservation Biology in the Department of Biology at Muhlenberg College, Allentown, Pennsylvania, USA. His book *Solid Air* was a splendid help to me in writing this book, in addition to his innumerable messages as a mentor.

Franz Hölker

My sincerest thanks to the brilliant Dr. Franz Hölker, head of the Research Group at the Leibniz Institute of Freshwater Ecology and Inland Fisheries (IGB), Berlin, Germany, for his excellent and inspiring research activity and invitation to deliver a lecture on "Interdisciplinary Eco-Architectural Darkness Design & Pragmatism" in Berlin in 2019. The research findings achieved by his group gave a special lift to this book, enriching it.

Kaj and Helena Öhman and Sakari Alhopuro

Many thanks to medical doctors Helena and Kaj Öhman and Sakari Alhopuro for their constant encouragement and support.

Special thanks to the professional staff of Hancock House Publishers Ltd; to editor Doreen Martens for her unbeatable skills in compiling my research material into a book, to Designer Jana Rade for her innate Lay Out skills, and to world-renowned Wildlife Biologist and Zoologist and Naturalist Myles Lamont for the assistance and coordination of this book.

FOREWORD

We are glad to introduce Julle Oksanen and his important book to readers. Having discussed the manuscript with Julle while it was still in progress, it is great to see it finished.

We're pleased to endorse a book this important to everyone interested in sustainable architecture and lighting, or biodiversity and birds. This treatise should be useful as well to other professions working with biodiversity, such as biologists, ecologists, and landscape architects. The built environment has a considerable impact on sustainability and the welfare of living organisms, and this book details the remarkably damaging effects of light pollution on biodiversity. The value of Julle's book is that it cleverly applies his technical knowledge about artificial lighting to fresh principles in architectural design. For our own work as architects, this book is what has been missing from the ranks of design guidelines.

A word about the two of us. Hannu Tikka is an acclaimed architect who has worked alongside Julle Oksanen as a design and education colleague, in completing built projects, and in organizing university workshops on integrated architectural and lighting design around the world. Professor and architect Antti Ahlava supervised Julle's doctoral dissertation and conducts the education and research unit Group X in the Architecture Department of Aalto University, where Julle was a Visiting Researcher in 2020-22, allowing him to concentrate on writing this book.

The publication follows from Julle's doctoral thesis, which was accepted in 2017 ("Design Concepts in Architectural Outdoor Lighting Design: Based on Metaphors as a Heuristic Tool"). Julle conceived the concept of "designing darkness" as a parallel to the way architects often sculpt light and space. One could here draw a straight line from Junichiro Tanizaki's aesthetic "In Praise of Shadows," through Venturi's and Scott Brown's nocturnal pragmatism in "Learning from Las Vegas," through to Julle Oksanen's environmental concept.

This follow-up manual, however, uses non-scientific language to describe the relationships between technical illumination and architecture, focusing on the impact of excessive artificial light—on birds, for example—and how to avoid doing damage with light. Julle is a person whose scientific work is deeply embedded in practice, so it's no wonder this volume contains numerous architectural examples. We don't know how much he is an ornithologist—he might be a "night owl," himself—but this book displays his justified concern for the future of our flying friends if their natural cycles and sense of navigation are disturbed. We understand that Julle's motivation was to compile a guidebook for architects, engineers and other designers that would help them prevent unnecessary bird deaths. But the tone of this volume isn't that of a doomsday sermon, but rather, encouraging, and enthusiastic.

Our long experience with architectural design projects in Finland and abroad leads us to recognize the need for guidelines on taking non-human life into account in design. Architects don't want to deliberately destroy nature, but misunderstandings or simple lack of knowledge sometimes lead to more damage than necessary. Lately, among architects and other designers, there has been a lot of discussion on post-Anthropocene philosophy- the idea that we are shifting into an era that stops emphasizing only human benefit. Increased focus on natural preservation, biomimicry, and a new interest in object-oriented ontologies have been some of the consequences of this "shift." Julle's book can be seen as an important contribution to the post-anthropocentric design discourse.

There might be other books, academic theses, and scientific articles on the impact of modern architecture on nature, but Julle's tome appears the first so far to apply technical knowledge about lighting and lighting installations to progressive architectural thinking.

—Antti Ahlava and Hannu Tikka

FROM THE AUTHOR

This book is primarily intended for people with a desire or opportunity to exert a bold influence on eliminating light pollution, in the interests of preserving the Earth's biodiversity: experts at universities and schools, organizations in various fields, suppliers, and decision-makers. Responses could take the form of action by various groups; independent initiatives and ideas, like this book; or by addressing a subject in an academic thesis. This book will also be helpful to anyone interested in learning more about the surprising deleterious effects of light pollution on biodiversity and how we can go about replacing energy-wasting, wildlife-harming outdoor lighting with more environmentally friendly Darkness Design.

In the following chapters, we'll take a deeper look at three paradigms: the Light Pollution Paradigm, the Technical Outdoor Lighting Paradigm, and the Paradigm for Design of Nocturnal Spaces, exploring their effects and offering proposals for their integration into Darkness Design.

This book was inspired by the research activities of Dr. Franz Hölker of the Leibniz Institute in Germany. In 2017, after I completed my doctoral dissertation at Aalto University, Finland, titled "Design Concepts in Architectural Outdoor Lighting Design, Based on Metaphors as a Heuristic Tool," I became familiar with Dr. Hölker and his findings on the deleterious effects of light on biodiversity. I was amazed and felt an enormous excitement, as I knew right away that this was an area in need of further research and potentially a book project where I could apply my education and over 40 years' experience in multifaceted lighting design. I visited Berlin to deliver a lecture at the institute on "Interdisciplinary Eco-architectural Darkness Design and Pragmatism." I had my motivation.

I began with the World Wildlife Federation's annual Living Planet Index (LPI), which tracks the populations of hundreds of species—birds, mammals, amphibians, reptiles, and fish—and measures changes occurring in the diversity of forests, oceans, and fresh water. The statistics showed indisputably that today's human consumption and behavior patterns are not sustainable. I was profoundly shocked by the diminishment in living species over the past 50 years. In 2022, the LPI (-68 %) shows that numbers of mammals, birds, fish, amphibians, and reptiles have fallen by 68% in the period 1970–2020. We are consuming more natural resources than our planet can sustainably produce. It has been estimated that by the middle of this century consumption by humans will double in relation to what the planet can support.

The WWF believes that governments can turn this trajectory in a more sustainable direction. This will necessitate more efficient production of goods and services, especially energy without fossil fuels. The energy-efficient development of technologies, buildings, and modes of transportation must continue. We also need to promote sustainable consumer habits to ensure diversity and productivity.

"We do not know exactly what the effects of exceeding the sustainability of Nature will be. It is therefore better to engage ourselves in taking things in a positive direction than to leave them to fate. "

— **Dr. Jonathan Loh, author of the Living Planet Report**

Darkness Design and Biodiversity grew out of my post-doctoral research conducted at Aalto University in response to what I have learned about this urgent issue. My hope is that this book will do its part to turn the Living Planet Index in a more sustainable direction by saving energy while building a future that welcomes darkness so Nature can flourish.

Helsinki Finland, September 2022
Julle Oksanen, PhD, Darkness designer

TABLE OF CONTENTS

DARKNESS DESIGN AND BIODIVERSITY: Introduction 1

Chapter 1: THE LIGHT POLLUTION PARADIGM 5

    Elements of the Light Pollution Paradigm 7
        Historical roots of the ecological crisis 8
        Human Resistance to Change 9
        General Education 11
        Capitalism from a Biodiversity Perspective 12
        Capitalism and Dependencies 12
        Biology Education. 13
        Light Pollution Research. 13
        Managing the Global Environmental Light Pollution Crisis. . . 15
        Ideal Darkness Design Direction 17
        Lighting Design and Light Pollution 17

Chapter 2: THE TECHNICAL OUTDOOR LIGHTING PARADIGM .21

    Introduction 22

    Upwards-Aimed Façade Lighting 24
        Illuminating space in vain 24
        Floodlights for facade lighting 30
        Interior lighting as facade lighting 31
        Technical lighting models for nocturnal architecture 32
        Legislative measures to promote bird-safe buildings. 34
        Effective methods to prevent bird collisions 35
        Lighting to avoid mirroring 40

    The "Light, Glass ('mirrors') and Bird Behavior graph 41
    Technical lighting examples for nocturnal architecture 49
    Orchestrated Darkness Design in a high-rise:
    Telenor Building, Oslo 53
    Other human structures and light pollution 62

Upwards-Aimed Tree Lighting 68
    Aura River tree lighting (as implemented in 1997) 68
    Tree lighting, molecular plant biology and light pollution. . . . 73
    Light pollution and the Aura River case study in 2021 76
    Heuristic mathematical calculus of light effect to the sky
    (and birds): 76
    The light veil's impact on biodiversity: lost night and disturbed
    migration. 80
    Biomimicry and biodiversity-friendly tree lighting solutions . . 83

**Upward-Aimed General Lighting: Large,
efficient light pollution producers** 95
    Illuminating space in vain 95
    Helsinki Airport Light Prisms (1997) 95
    Light-pollution free "Ice Boulders" (a renovation project and
    model for new projects) 97

Road/Street/Vehicle Lighting 99
    Lack of cooperation and a great opportunity 99
    Obstacles to cooperation, and some solutions 101
    A legacy of roadway lighting. 105
    Road lighting today 106
    Impact of current road lighting on biodiversity. 109
    Road lighting, clouds, and light pollution 117
    Modern vehicle light systems + intelligent LED road lighting 120
    Summary: Road, Street, and Vehicle Lighting 126

Advertising Lighting 127
    Billboards 127
    Other advertisement lighting solutions 128

ULOR-CT (Upward Light Output Ratio – City Total) 129
    ULR and ULOR 129

ULOR-CT 135

Changing the Process 140

Chapter 3: PARADIGM FOR DESIGN OF NOCTURNAL SPACES . 143

Introduction 144

Biodiversity and Technical Tools for Darkness Design 146
  Modern technical tools for Darkness Design 147
  Global growth of intelligent lighting 156
  Opportunity and Challenge 156

Biodiversity and architectural Darkness Design tools 157
  The Darkness Design process 157
  Darkness Design master plan 158
  Darkness Design details 158
  The design process according to Richard Kelly 159
  Hopkinson's Scale of Apparent Brightness 167
  Heuristics and pragmatism as instruments 169
  Practical use of Hopkinson's Scale 170

Heuristic approach 176
  Approaching in darkness 177
  Sample computer calculations 181

**Nocturnal Darkness Design, creativity structures
and biodiversity** 186
  About shadows and nocturnal darkness 186
  Shadow and darkness composition 193
  New technical tools for designing darkness 197
  Smart Cities and lighting control systems 198
  Integrating facade lighting into urban intelligent
  control systems 201

Nocturnal Master Plan strategies for cities 203
  "Shadow is light's best friend" 203
  Step 1: Analyzing existing lighting 204
  Step 2: Character zones and defining the testing area 205

Step 3: Circulation and the testing area 206
Space, Darkness Design, traffic LED lighting & glare 207
A partial solution in the testing area: intelligent
LED headlights 210
Employing Hopkinson's Scale in the testing area 211
LED solutions in the test area: large-surface lighting 212
Step 4: Places of night interest in the testing area 214
Don't test façade lighting 215
Do test: interior lighting as facade lighting 216
Other interesting elements in the testing area 217
Step 5: After testing, implementation 220

Our Future 221

**Appendix 1: Telenor: Orchestrated Darkness Design
in a high-rise building** 223
Darkness Design concept 225
Heuristic metaphors 226
Darkness Design master plan 229
Lighting design details 244

Demonstrations and implementation 246
Final thoughts on Telenor 246

Appendix 2: Futuristic "City 2030" 247
Ambient lighting in the City 2030 structure 248
Structure of "City 2030" 249
Luminaires used in calculation 251
Luminaires and basic calculation values 252
Calculations for basic values without Brando luminaires . . . 254
Calculations of basic values with Brando luminaires 256
Creating the public "living room" 258
Heuristic energy-saving calculations using "rule of thumb" . 263

About the Author 269

DARKNESS DESIGN AND BIODIVERSITY: INTRODUCTION

We humans have put plenty of effort and money into studying external planets and solar systems, while not doing enough to take care of our own planet. Climate change is a scientific fact. One significant factor in our violation of the natural world is light pollution, which is contributing to destroying Earth's biodiversity at an accelerating rate.

Unnecessary and ill-designed artificial light at night (ALAN) disrupts the natural cycles of organisms, with largely unforeseen consequences for their biology and interaction in the ecosystem. ALAN causes disorientation, attraction or repulsion from the illuminated area. This light causes daily or seasonal transitions, problems in feeding, communication problems, problems in reproduction and bird collisions with buildings (an estimated one billion birds die in these collisions each year in North America).

We often illuminate space and the sky for no reason, and because "light doesn't disappear," part of this useless light reflects back to the surface of Earth, penetrating the seas and disturbing the behavior of fish and benthic animals. A 2017 *Science Advances* article by Kyba et al., titled "Artificially lit surface of Earth at night increasing in radiance and extent," describes evidence from the calibrated Suomi NPP satellite radiometer, which was designed for night light research 833 kilometers above the Earth's surface. The evidence shows that between 2012 and 2016 alone, global light pollution grew by 2.2% per year. And this continues. We are losing the night; stars are disappearing and energy consumption increasing.

The changed behaviors of various animal species caused by this unnatural light, and a reasoned response to this issue by altering nocturnal design methods, are described in the Technical Outdoor Lighting Paradigm chapter of this book. How natural patterns of light have changed in our environment over the past century is explored in the chapter on the Light Pollution Paradigm. This is attributable to many independent design elements, including excessive street and road lighting, advertising lighting, unprofessional architectural lighting, technical lighting solutions, security lighting, building illuminations inside and outside, tree lighting, show lighting, park lighting, color lighting, etc. Satellite measurements prove that no effective actions have been taken thus far to resolve this. No biological research on the effects of light pollution on biodiversity has been undertaken in relation to nocturnal architectural design. This research is much needed and should be incorporated into lighting design recommendations and practical projects worldwide.

This book provides some practical proposals for changes in lighting design that could be utilized in multidisciplinary international collaboration models, in response to the findings of 21 research reports exploring the destructive effects of light on biodiversity.

Eliminating the problem also requires a clear, well-managed structure of international cooperation. In 2021, the *International Journal of Environmental Research and Public Health* published an encouraging analysis titled "Urban Lighting Research Transdisciplinary Framework – A Collaborative Process with Lighting Professionals." Since light pollution is a collective, multidisciplinary problem of humanity that needs to be solved together, and since there are some 40 active environmental research institutes in the world, it makes sense for an international umbrella organization such as the United Nations to lead the project (see more in Chapter 1).

Lighting design, as a profession in its own right, needs to be more professionally integrated with architecture, using appropriate nocturnal design methods dubbed Darkness Design. These proposed methods are introduced in Chapter 3, Paradigm for Design of Nocturnal Spaces. A global problem is the dire lack of professional training in nocturnal design in all lighting sectors, electrical engineering, architecture, landscape design, manufacturing, biology, energy production, light leadership education, city planning organizations, lighting implementation stages, etc. Decades of development in the various fields mentioned above has seen lighting design siloed into sub-sectors such as road lighting, vehicle lighting, landscape lighting, show lighting, technical lighting, architectural lighting, and theatrical lighting. In some cases, there are inherent conflicts, for example developments in road lighting vs. vehicle lighting, technical lighting design vs. architectural lighting design, etc.

This book explores the possibility of eliminating light pollution with new nocturnal design guidelines that result from cross-disciplinary and integrated research. We examine three paradigms: the Light Pollution Paradigm, the Technical Outdoor Lighting Paradigm, and the Paradigm for Design of Nocturnal Spaces. This combination, viewed from an unprejudiced and professional perspective, leads us to design methods that are are free of light pollution— "Darkness Design Recommendations for Nocturnal Architecture."

Protecting Earth's biodiversity while allowing users to safely enjoy the nocturnal space requires the designer to "orchestrate darkness," instead of spreading excessive light. The Telenor project presented in this book demonstrates one example of how to do this successfully.[1]

---

[1] Given the massive lack of education in architectural lighting design worldwide and the total lack of education in darkness design, it would be helpful for serious readers of this book to also delve into my dissertation, "Design Concepts in Architectural Outdoor Lighting Design, Based on Metaphors as a Heuristic Tool": http://urn.fi/URN:ISBN:978-952-60-7391-0

CHAPTER 1: THE LIGHT POLLUTION PARADIGM

We humans have caused an alarmingly large degree of damage to the Earth and its biodiversity, reshaping its surface and what lies beneath. Nature provides us with oxygen, water and food. If we are to survive, the ecosystems must be saved.

The plants and animals we live with are doing their best to adapt to living within the fragmented stretches of natural environment our infrastructure has left for them. Over the past 50 years, our actions have already destroyed 70 per cent of the life contained in the "Living Planet Index."

"Each one of us must do our part to save Nature" may sound like a cliché. But in 2017, I decided to do my own bit: to work at eliminating light pollution from architecture. I began by exploring the Light Pollution Paradigm, a way of charting the human factors in the problem of light pollution, by plotting them on horizontal and vertical axes. The vertical axis indicates the human response to light pollution and other biodiversity disasters, while the horizontal axis represents human attitudes toward eliminating light pollution and conserving biological diversity.

Exploring various elements of this paradigm was thrilling and thought-provoking—for example, the roots of history and biodiversity, the effects of capitalism on biodiversity, the inherent reluctance of humans to cooperate, their resistance to new things, educational structures, rigid and massive decision-making structures, and the influence of time on all of this.

My observations on this issue motivated me to write this book and helped lead me to ideas and possibly even solutions to the problems revealed in this paradigm.

I hope this book will encourage students in many different fields—philosophy, architecture, interior design, branches of engineering, vocationally oriented students, students of lighting, etc.—to find related topics for their own theses or final projects and boldly gain entry to a new world. The doors are open. Go for it!

ELEMENTS OF THE LIGHT POLLUTION PARADIGM

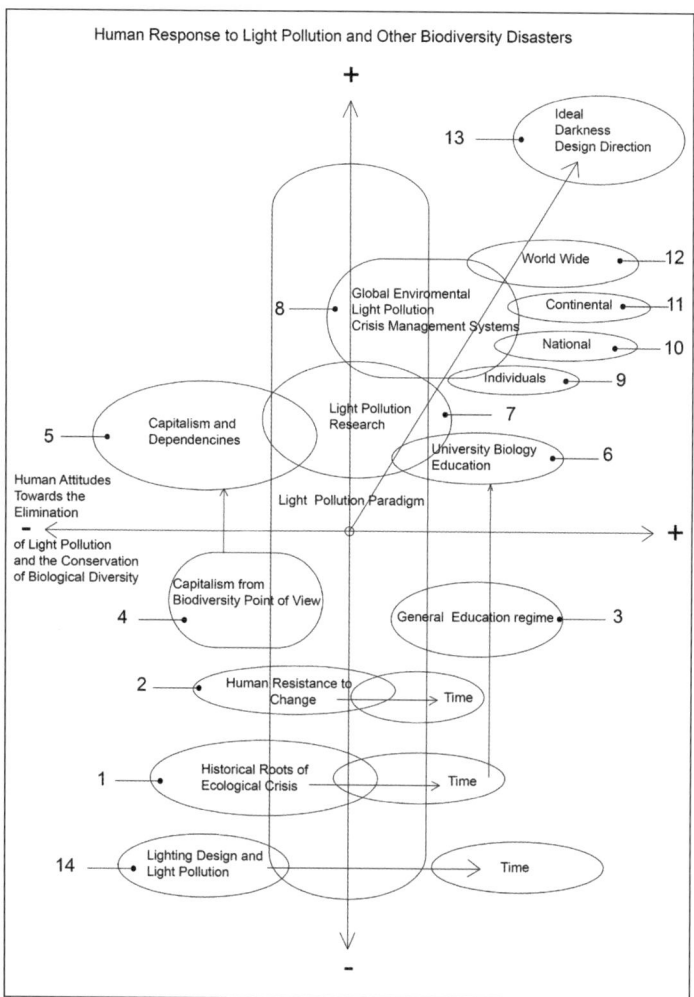

Figure 1: The Light Pollution Paradigm is placed on the axes "Human Response to light pollution & other biodiversity disasters" and "Human attitudes towards the elimination of light pollution & the conservation of biological diversity". Elements that influence the paradigm (numbered 1-14) are the basic factors explored in this book, which offers an opportunity for a paradigm shift in a biodiversity-saving direction, substantial energy savings, and a whole new era of nocturnal architecture.

Historical roots of the ecological crisis

American historian Lynn Townsend White, in a well-reasoned examination of the roots of our ecological crisis in the journal *Science*[2], cites the prevalence of certain elements of traditional Christian dogma as a prime cause. He asserts that the crisis will continue "until we reject the Christian axiom that nature has no reason for existence save to serve man."

Some interpretations assert that God gave humans the Bible as a written scripture, while Nature was conceived primarily as a symbolic system through which God speaks to human beings. The religious study of nature to better understand God was long described as "natural theology."

The first steps in a scientific direction came in the early 13th century, with scholars attempting to decode the physical symbols of how God communicates with humans. The rainbow, for example, was no longer simply a symbol of hope first sent to Noah after the Deluge. Early researchers Robert Grosseteste, Friar Roger Bacon, and Theodoric of Freiberg produced sophisticated research on rainbow optics, doing so, in the atmosphere of those times, as a "project of religious understanding."

The 18th century Enlightenment in Europe emphasized the importance of reason and knowledge in pursuing fundamental reforms in culture and social life. The liberalism of the Enlightenment provided the framework for revolutions in the United States and France, the rise of capitalism, and the birth of socialism. During the Enlightenment, reason replaced mysticism and science replaced beliefs. A free-market economy replaced trade policy, and the arbitrariness of the State was replaced by human rights.

In the world of the Enlightenment, the role of nature was emphatically instrumental. With the help of science and technology, it was imagined that humans could control and exploit nature.

The term *ecology* first popped up in the English language in 1873, Ecology, as we define it now, is a multidisciplinary branch of biology that studies how organisms and species interact with each other and the inanimate environment around them.

The crystallization of this then-novel concept within an unprecedented situation in the 19th century marked the first small step towards understanding the importance of conserving biodiversity. But there were no measures in place to combat biodiversity disasters.

---

[2] March 10, 1967, Vol. 155, No. 3767 / "Historical Roots of Our Ecological Crisis"

Combined with the greed that is said to drive modern people to want "20% more of everything" all the time, this view of our relationship with the rest of the natural world has led to a serious upset in the balance of nature. Only now, in the 21st century, have we truly realized how our behavior has damaged the biodiversity of our planet. With the influence of time, the Historical Roots of the Ecological Crisis element has shifted toward the positive side of the Light Pollution Paradigm.

Implementing a paradigm shift in light pollution will require professional collaboration among architects, engineers, and biologists, clear and broadly researched arguments, and guidance toward eliminating light pollution in architecture and achieving suitable design in our nocturnal spaces.

Human Resistance to Change

Rosabeth Moss Kanter, Professor of Economics at Harvard University, described in an article "Ten Reasons People Resist Change."[3] I've employed and modified some of these to serve this book. Seven reasons that apply to this subject:

Loss of control. When it comes to redesigning lighting, biodiversity-friendly design needs to move from lighting design to skillful, shadowy nocturnal design—Darkness Design. This is a change that will be exceptionally disruptive for those who specialize in the field of lighting. It's not a question of who wields power, but rather whether we can put aside our concern for professional autonomy to take preserving biodiversity into account. Nature can't speak for itself. The purpose of this book is to broaden the possibilities of Artificial Light At Night (ALAN) design to assist skilled designers, so we can all enjoy glare-free and contrast-rich nocturnal spaces that avoid disturbance to other species.

Excess uncertainty. People are often more willing to languish in misery than to venture into the unknown. In biodiversity-conscious lighting design methods, transitioning to shadow-rich nocturnal space design, Darkness Design, is hard work and requires strong motivation. Overcoming inertia requires a sense of security and an inspiring vision. This book will attempt to create process certainty with clear arguments that are somewhat new to the profession.

---

3   Harvard Business Review September 25, 2012

Surprise, surprise! Decisions suddenly launched on people are usually met with opposition, with no time to get used to the idea or prepare for the consequences. It is always easier to say no than to say yes.

The purpose of this book is to motivate experts in the field to make major changes in design methods and product manufacturing that will minimize light pollution in old environments and eliminate it in new projects. Unfortunately, there is no time to wait for a gradual process of motivation. Taking biodiversity into account in lighting/darkness recommendations, based on purely technical arguments, may lead to fear, a strong "group selfishness" response and refusal to cooperate.

But recent irrefutable research compels us to include effects on biodiversity in our design considerations. This book evaluates both the rationale for and possibilities of new criteria for nocturnal, shadow-rich design that accommodates the needs of natural species. Adopting these new criteria in future lighting design recommendations will require a positive attitude and seamless collaboration among architects, engineers, and biologists.

Everything seems different. Change is meant to bring something different, but *how* different? As ordinary beings, we tend to live by routines; changes push us to become more conscious, sometimes uncomfortably so. Too many differences can be distracting or confusing. The transition from lighting design to shadow-rich nocturnal space design involves new and perhaps confusing arguments. But research findings from professional biologists and institutions such as the Leibniz Institute point clearly to the need to eliminate or at least minimize light pollution.

Loss of face. Change is, by definition, a departure from the past. Those who subscribed to the latest version are likely to defend it. When there is a major change in strategic direction, those responsible for the previous direction fear the perception that they were wrong. But the current paradigm has been in place in the lighting industry for too long. The biggest issue may be that even the umbrella organization for basic research in the field, the CIE (International Illumination Commission), has not cooperated at all with biologists, who have been studying light pollution problems for decades. Understanding that the world has changed will make it easier to let go and move forward, enjoying new challenges.

Concerns about competence. Sometimes there's opposition to change because it makes people feel powerless. For example, lighting designers may express skepticism as to whether new shadow-rich architectural nocturnal space design methods work. But the unspoken worry may be that their skills will become obsolete. A comprehensive reorganization of design methods must provide a wealth of collaboration, information, training, mentors, and support systems. New design methods will be based in part on existing lighting design arguments. Concepts, master plans, detailed design, computer programs, technical mathematical formulas, and the results of technical research—these can all be modified to suit the new design methods. These factors facilitate change and help and promote collaboration among biologists, architects, and engineers.

More work. It's a universal challenge: changing from excessive illumination back to reasonable darkness will indeed take a lot of work and disruption, which can be overwhelming for those closest to the issue. A decades-long replacement of the massively increased sources of lighting pollution with professional, environmentally friendly Darkness Design methods will require a huge amount of work from the entire lighting field.

General Education

King's School, in Canterbury, England, was founded in 597, making it the oldest school in the world.

Formalized education has been with us for many centuries since then and become universal internationally, creating a framework for people to solve problems and accomplish tasks together. Globalized education has helped us understand common problems and develop various solutions. Preserving biodiversity has become a clear, global collective project, as the significance of its destruction has become broadly understood. An excellent example of the influence of education is Sweden's Greta Thunberg (b. 2003), who has won international awards and fame as the founder of an international climate movement for schoolchildren.

Capitalism from a Biodiversity Perspective

Capitalism, which has its roots in the Enlightenment, emphasizes profit and competition. The human goal of attaining "20% more everything" (author) feeds the growth of capitalism and has negative side effects for those indirectly affected. For example, greenhouse gas emissions affect the common climate, no matter where they occur. However, no state has sufficient incentives to achieve the reduction in emissions required, and all states expect others to shoulder their responsibilities without doing much of anything themselves. The same philosophy applies to reducing light pollution, although pollution of the atmosphere by light mainly concerns land masses and not the whole Earth.

Capitalism and Dependencies

We're now at the point where there's broad recognition that we as a species are dependent on Nature. Global ecological crises and related high-quality education and research, as well as multifaceted international cooperation, have put pressure on capitalist economies to change their thinking. Capitalism isn't about to disappear, but it is revising its essence. Every state wants to be seen to be involved in developing and selling materials and methods for dealing with ecological crises. Previously, financial returns were maximized by producing energy using the cheapest and most polluting materials (coal and oil).

Revenues are now sought through developing wind power and solar energy.

To deal with light pollution, international cooperation is being developed, for example using costly satellite systems to measure light pollution. Dutch scientists have developed GPS sensors that are mounted on the backs of bees to study their routes and the reasons for their disappearance. The Leibniz Institute studies the multifaceted effects of light pollution on biodiversity. All such projects in the field pave the way for a new form of capitalism. Maybe we could name it Biodiversity Capitalism. In any case, capitalism's dependence on dealing with new biological challenges has allowed it to play its part in supporting the human response to light pollution and other biodiversity disasters, and to elevate the Light Pollution Paradigm in a positive direction in the diagram.

Biology Education

Biological education consists of research in both biological disciplines and general disciplines (physics, mathematics, chemistry, political economy, history, and philosophy, and so on). Biology itself is divided into several areas: ecology, microbiology, physiology, genetics, biochemistry, biophysics, medicine, agriculture, forestry, fisheries, and game economics. The fastest-growing discipline is molecular biology. Some biology disciplines have incorporated physics, chemistry, and mathematical and statistical methods. Biology disciplines interact with each other and often with related sciences.

A good example is the Leibniz Institute's studies on light pollution, the results of which this book presents as arguments in favor of rich nocturnal design, free from light pollution.

Biology diplomas and certificate programs offered in Europe include bachelor's degrees in 119 institutions, master's degrees in 140 institutions, and doctoral degrees in 54 institutions. The USA reports 29,278 degrees awarded in Biological & Physical Sciences in 2019 alone (associate, bachelor's, master's, and doctoral degrees). However, light-pollution control applications are completely lacking in these programs.

Light Pollution Research

This book is inspired by readings of light pollution studies by professional researchers and professors at various universities and research institutes, such as the Leibniz Institute. Below is a brief list of these reports and their impact on biodiversity.

"Loss of Night–Transdisciplinary Research on Light Pollution" (Dr. Franz Hölker and Dr. Klement Tockner, Leibniz Institute of Freshwater Ecology and Inland Fisheries, 2010)

"The Biological impacts of artificial light at night: The Research Challenge" The Royal Society Publishing (research in 2015).

"Artificial Light at Night Affects Organism Flux across Ecosystem Boundaries and Drives Community Structure in the Recipient Ecosystem" *Frontiers in Environmental Science* (research in 2017)

"Artificial night lighting inhibits feeding in moths" The Royal Society Publishing (research in 2017)

"Artificial light at night decreases biomass and alters community composition of benthic primary producers in a sub-alpine stream" *Limnology and Oceanography* (research in 2017)

"Artificial lit surface of Earth at night increasing in radiance and extent" *Science Advances*/Research Article (research in 2017)

"Insights into the Social Behavior of Surface and Cave-Dwelling Fish (Poecilia Mexicana) in Light and Darkness through the Use of a Biomimetic Robot" *Frontiers in Robotics and AI* (research in 2018)

"Long-Term Comparison of Attraction of Flying Insects to Streetlights after the Transition from Traditional Light Sources to Light Emitting Diodes in Urban and Peri-Urban Settings" *Sustainability* (research in 2019)

"Observing the Impact of WWF Earth Hour on Urban Light Pollution: A Case Study in Berlin 2018 Using Differential Photometry" *Sustainability* (research in 2019)

"Remote Sensing of Night Lights—Beyond DMSP" *Remote Sensing* (research in 2019)

"Snowglow: The Amplification of Skyglow by Snow and Clouds Can Exceed Full Moon Illuminance in Suburban Areas" *Journal of Imaging* (research in 2019)

"Artificial Light at Night Influences Clock-Gene Expression, Activity, and Fecundity in the Mosquito Culex pipiens f. molestus" *Sustainability* (research in 2019)

"Light Pollution, Circadian Photoreception, Melatonin in Vertebrates" *Sustainability* (research in 2019)

"Impacts of artificial illumination on the development of a leafmining moths in urban trees" *International Journal of Suitable Lighting* (research in 2020)

"Avian circadian organization: A chorus of clocks" (*Frontiers in Neuroendocrinology* 35 (2014) pp. 76-88

"Entrainment of the Circadian Activity Rhythm to the Light Cycle: Effective Light Intensity for a Zeitgeber in the Retinal Degenerate C3H Mouse and the Normal C57BL Mouse" *Physiology & Behaviour*, Vol. 24 pp.523-527)

"High-intensity urban light installation dramatically alters nocturnal bird migration" *PNAS*, October 17, 2017, vol. 114)

"Light, Glass, and Bird-Building Collisions in an Urban Park" *Northeastern Naturalist*, vol. 22 No. 1

"Avian Window Strike Mortality at an Urban Office Building" *The Kingbird* 2006 September 56 (3)

"Bird Safe Glass Legislation in North America May 26, 2020"[4]

"Effective And Attractive Ways To Make Building Windows Safe For Birds" Bryan Lenz, American Bird Conservancy, February 04, 2020

Solid Air Invisible killer: Saving Billions of Birds from Windows Dr. Daniel Klem. Jr. (ISBN 978-0-88839-646-4)

---

4 https://www.walkerglass.com

Managing the Global Environmental Light Pollution Crisis

There are about 40 environmental research institutes in the world. These organizations undertake research on sustainable management of resources, including water, energy, and biodiversity. Many have released information about light pollution mainly in articles and press releases, but solutions for avoiding it are hard to find.

UNEP: United Nations Environment Program

Figure: UNEP

UNEP, the United Nations Environment Program, is the UN's leading environmental authority. UNEP uses its expertise to strengthen environmental standards and practices while helping countries implement environmental obligations at the national, regional and global level. UNEP's mission is to provide leadership and encourage partnership in caring for the environment by inspiring, informing, and enabling nations and peoples to improve their quality of life without compromising that of future generations.

Six Areas of Concentration

UNEP has organized its work into six strategic areas, guided by scientific evidence, the UNEP mandate and priorities emerging from global and regional forums.

1. **Climate change**: UNEP works to strengthen the ability of countries to integrate climate change responses by providing leadership in adaptation, mitigation, technology and finance. UNEP is focusing on facilitating the transition to low-carbon societies, improving understanding of climate science, facilitating development of renewable energy, and raising public awareness.

2. **Post-conflict and disaster management**: UNEP conducts environmental assessments in crisis-affected countries and provides guidance for implementing legal and institutional frameworks to improve environmental management. Activities undertaken by the Post-Conflict & Disaster Management Branch currently include environmental assessments in Afghanistan, Côte d'Ivoire, Lebanon, Nigeria and Sudan.

3. **Ecosystem management:** UNEP facilitates management and restoration of ecosystems in a manner consistent with sustainable development and promotes use of ecosystem services. Examples include the Global Programme of Action (GPA) for the Protection of the Marine Environment from Land-Based Activities.

4. **Environmental governance**: UNEP supports governments in establishing, implementing and strengthening their processes, institutions, laws, policies and programs to achieve sustainable development at the national, regional and global levels, and mainstreaming the environment in development planning.

5. **Harmful substances**: UNEP strives to minimize the impact of harmful substances and hazardous waste on the environment and human beings. UNEP has launched negotiations for a global agreement on mercury and implements projects on mercury and the Strategic Approach to International Chemicals Management (SAICM) to reduce risks to human health and the environment.

6. **Resource efficiency/sustainable consumption and production**: UNEP focuses on regional and global efforts to ensure natural resources are produced, processed, and consumed in a more environmentally friendly way.

Where light pollution fits

UNEP points 3 and 4 (ecosystem management and environmental governance) would be great points from which to coordinate efforts to eradicate light pollution, an effort that is only just beginning in all international organizations. UNEP's websites yield only articles and reports about light pollution, presenting no efficient scientifically based solutions. The goal of this book is to propose solutions for closer cooperation between scientists who understand lighting and researchers who understand biological problems. There is no time to waste.

Ideal Darkness Design Direction

To end light pollution, we need to maximize the human response to light pollution and other biodiversity disasters (vertical axis on the chart) and also human attitudes toward eliminating light pollution and preserving biological diversity (horizontal axis). This will lead to a much-needed paradigm shift: the creation of biologically protective Darkness Design to replace technically oriented lighting design.

Lighting Design and Light Pollution

This element of the current paradigm reveals two problems: lack of cooperation between actors in the sector, and indifference to the problems caused by light pollution.

Lighting Design

Architectural outdoor lighting design is currently based on a systemized, technical approach that adheres to certain rules and is largely impervious to external feedback and creative aesthetic development. Creative, aesthetically pleasing architectural lighting design is the first step toward eliminating light pollution from architecture.

Technical lighting has a legacy of more than 100 years. There are hundreds of technical lighting labs and research centers around the globe. Research and lighting design recommendations issued by the umbrella organization, CIE, serve illumination engineering societies in more than 250 countries, and those recommendations are followed by thousands of trained engineers on their projects.

The recent development of incredibly versatile intelligent lighting systems—motivated mainly by energy savings, for example for LED road lighting—marks a masterful performance in technical lighting. When used correctly and applied to biodiversity-respecting, shadow-rich nocturnal design methods, these versatile intelligent lighting systems are an ideal means of eliminating light pollution.

This book presents proposals and examples. At the same time, it offers people working in the field of technical lighting with great motivation to cooperate with biologists.

The status quo in architectural lighting

It's been estimated that more than 95% of the people designing lighting projects are self-taught lighting designers. Most universities offer only a two-hour course in lighting design, and then only under the rubric of "building systems."[5] My subjective knowledge and 40 years of international activities in the field of lighting education and design suggest that the global situation is the same now as in 1987, with rare exceptions. For years, the International Association of Lighting Designers (IALD) has worked hard with universities to make lighting design an integral part of degree programs in architecture.[6] Unfortunately, in the USA, for example, in many architecture schools, lighting design is still integrated into courses on environment control systems, where electricity and acoustics play a greater part than light and lighting design.

The status quo in technical lighting

Thousands of pages have been written and much ink expended on arguing technical points in lighting. This is an outcome of strenuous efforts by technically oriented people. In 1900, the predecessor of the Commission Internationale de L'eclairage (CIE, i.e. International Commission on Illumination) was founded to research and standardize the properties of gas lighting. Standardization took a major step forward in 1931, when the CIE introduced an international trichromatic colorimetry and photometry system, known as the CIE System, still largely in use today. Engineers became active in illumination engineering societies.

Sadly, the more visual skills of the lighting designer declined. Lighting design work shifted from the hands of visually oriented people to the hands of technically oriented people. Engineers have accomplished much as far as the quantity and distribution of light are concerned, but lighting design is often unbalanced and skewed. Because of this, projects look technical and often suffer from a lack of aesthetic merit.

It's useful to look at and study lighting with fresh eyes. The study of lighting requires an open mind unfettered by the judgments of earlier actions, solutions or recommendations.

Studying existing solutions carefully, along with their connection to technical lighting recommendations, points the way to better Darkness Design solutions. Lighting design is largely implemented worldwide by electrical engineers, sometimes in collaboration with an architect

---

5   Gad Giladi, 1987. "A Research in The Studies of Lighting Design in Schools of Architecture around the World" Master´s thesis, Parsons School of Design, New York, New York.
6   http://www.iald.org/trust/OutreachtoStudentsofArchitecture.asp

or other design-oriented professional. There is also a small group of professional lighting designers who are self-educated, without institutional credentials, and are pioneering lighting design in parallel with the rapid development of new technologies, such as RGB LED systems. This creates a risk of unbalanced lighting solutions that may have a "wow" factor but look odd, offering no answer to questions such as, "Why color?" In the wrong environmental context, these result in unfortunate nighttime architecture solutions.

Time

Time also has been a factor in this paradigm. Architectural lighting design has been guided by major technology changes. Technical revolutions have been so pervasive as to divert attention from the basics of lighting design. At the moment, LED, OLED, electronic control, and their rapid development are the focal point. Development in architectural lighting design, not to mention shadow-rich, professional nocturnal design—Darkness Design—has been minimal. This means there is a modest level of knowledge and understanding throughout the field among engineers, architects, and other lighting-industry actors. The gap between architectural development and lighting design development is dire indeed, and this situation must be remedied. Universities need to find more space in their architecture programs to change the world of architectural lighting/Darkness Design. Light is still the fourth dimension of architecture and is screaming out for skilled darkness designers and the creation of shadow-rich nocturnal design spaces.

Light Pollution

Removing light pollution from existing and future installations is a demanding process. I hope that this book will provide a fruitful foundation, as well as strength, enthusiasm, and motivation to intervene in the process of eliminating light pollution, which requires cooperation and strong collective education with various interest groups, international organizations, ministries, and industry actors.

CHAPTER 2: THE TECHNICAL OUTDOOR LIGHTING PARADIGM

INTRODUCTION

Like music, light is a universal language. Which makes it lamentable that there is such a lack of comprehensive education in the field of lighting, leading to an overall lack of expertise and inability to cooperate among experts in related fields.

However, the applied research of engineers seeking practical technical solutions is beginning to bridge the gap, yielding new opportunities for collaboration between various sectors. Examples include compensating for road lighting through the intelligent vehicle lights of cars (as described in John Bullough's article "Will road lighting wither?" introduced later in this book), or the management of various suburbs or indeed an entire city through automated holistic lighting with pre-set control systems.

The design of such systems should enlist not only electrical and electronics specialists but also biologists, experts in architectural lighting, architects, landscape gardeners, specialists in vehicle lights, representatives of systems suppliers, experts with the party commissioning the work, etc. Professionals accustomed to independent decision-making may find such a notion of broad cooperation strange. The complexities of this are addressed in greater detail in Paradigm 3, "Paradigm for Design of Nocturnal Spaces."

Eliminating light pollution to preserve biodiversity entails knowledge not only about the techniques of outdoor lighting but also the devastation or at least disruption in the natural world caused by its misuse. This chapter will introduce various practical lighting design projects to show how the right kind of Darkness Design can modify and at least partially redeem traditional projects that are otherwise detrimental to biodiversity, and in the process achieve architectonically interesting results.

Tremendous advances in lighting technology and their creative use can lead to the total elimination of light pollution and astounding energy savings (up to 80%) at the same time. The Telenor Headquarters building at Fornebu in Norway is a comprehensive example of darkness orchestration. Ingenious architect and artist Vesa Honkonen and the author were awarded the international Norsk Lyspris 2003 prize for their success with this project.

The more disciplines included on the design team and the more deeply the design is explored, the better the results, maximizing the visual delight of a given project. Ideal orchestration of the Darkness Design team will be achieved when all the elements needed for the project are under optimal control. Depending on the project, sub-fields include: biology, architecture, lighting technology, electrical design, spatial expertise, control systems for electronic lighting, management of the future of vehicle lights, and the experts representing the supplier and the commissioning party. That's the perfect orchestra for Darkness Design—and the conductor needs to be some global organization.

In this chapter, we'll look at elements (1-13) of the Technical Outdoor Lighting Paradigm with the help of real-life examples. The proposed changes for each element are presented as a basis for decision-makers and are also integrated in the chapter titled Paradigm for Design of Nocturnal Spaces. The goal is to maximize the length of the arrow labelled "Ideal outdoor lighting paradigm direction," taking into account all aspects of the changing process element 14: "New Thinking", "New Lighting Technology", and "Collaboration".

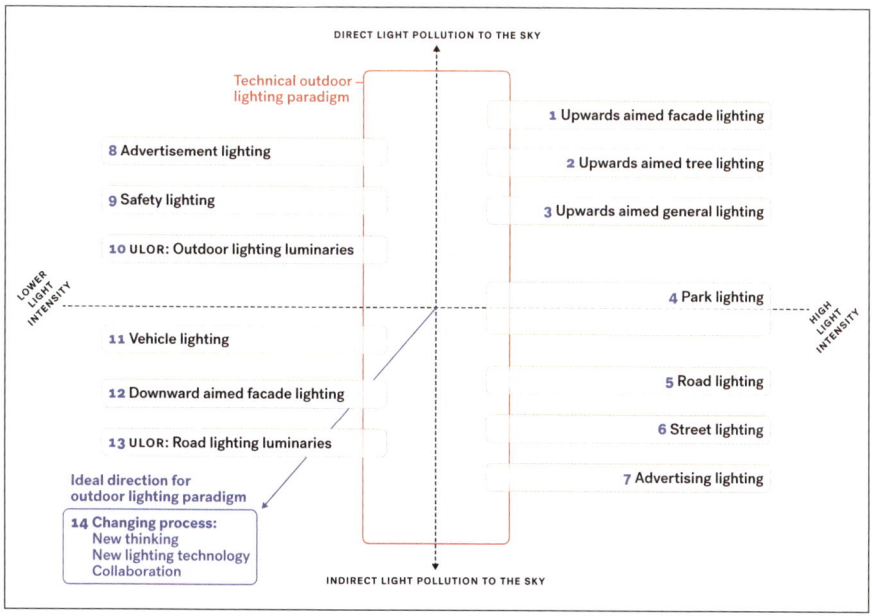

Figure 2: Elements of the Technical Outdoor Lighting Paradigm are plotted here on two axes representing (vertical) Direct Light Pollution to the Sky vs. Indirect Light Pollution to the Sky, and (horizontal) Lower Light Intensity and High Light Intensity. Explanations and treatment of each element appear below.

UPWARDS-AIMED FAÇADE LIGHTING

Illuminating space in vain

Improvements in the efficiency and brightness of light sources, in tandem with growing GDP, have resulted in tremendous increases in artificial light use over the past century, according to a *Science Advances* article titled "Artificial lit surface of Earth at night increasing in radiance and extent." Historically, lighting has experienced a strong rebound effect: increases in luminous efficacy result in greater use of light rather than energy savings. Humans in every time and place have tended to use as much artificial light as they can buy for about 0.7% of GDP. Outdoor lighting became commonplace with the introduction of electric light and grew at 3% to 6% per year in the second half of the 20th century. As a result, the world has experienced widespread "loss of night," with half of Europe and a quarter of North America experiencing substantially modified light-dark cycles.

The Visible Infrared Imaging Radiometer Suite Day-Night Band (VIIRS DNB) instrument came online just as outdoor use of light-emitting diode (LED) lighting began in earnest. This sensor provides the first-ever global calibrated nighttime radiance measurements in a spectral band of 500 to 900 nm, which is close to the visible band, with a much higher radiometric sensitivity than possible before, and at an improved spatial resolution of near 750 m (15). This allows scientists to make measurements of change in lighting at neighborhood-scale (rather than city or national scale) for the first time.

Cloud-free data shows that over the period 2012–2016, both the lit area and the radiance of previously lit areas increased in most countries in the 500–900 nm range, with global increases of 2.2% per year for lit area and 2.2% per year for the brightness of continuously lit areas. The numbers were stable in some of the world's brightest countries (for example, Italy, the Netherlands, Spain, and the United States). With few exceptions, growth in lighting occurred throughout South America, Africa, and Asia.

On one hand, the increased luminous efficacy of LEDs decreases energy consumption. On the other hand, even though nearly all new outdoor lighting installations now make use of LEDs, increased lighting necessarily implies new energy consumption. For this reason, increases in observed radiance are nearly certain to be due to increases in installed visible light and, therefore, also mean raised energy consumption.

Figure 3: The recently calibrated Suomi NPP satellite radiometer, designed for night light research, 833 km above the Earth's surface, has shown that between 2012 and 2016, light pollution grew by 2.2% per year, and continues to do so. We have lost the night, stars are disappearing, and energy consumption is increasing. Credit: NASA

Figure 4: Energy-saving LED increases Light Pollution. Credit: NASA

We illuminate space in vain because light doesn't disappear; some of that useless light gets reflected back to Earth by the atmosphere, penetrating the oceans and disturbing the behavior of sea life. According to one research article[7], "In aquatic environments, the impacts of light pollution from ALAN on the biomass and composition of primary producers in benthic communities have been found. Microbial diversity and respiration are also altered under the influence of light pollution. Zooplankton, such as daphnids, exhibit changes in their daily vertical migration pattern in response to light pollution at the very low light levels produced by skyglow. Drift patterns of aquatic insect larvae are likewise modified by the light pollution level at ~1 lx. In sum, this may lead to general changes in food web interactions and ecosystem functions."

---

7   Bruning A, Kloas W, Preuer T, Hölker F (2018) "Influence of artificially induced light pollution on the hormone system of two common fish species, perch and roach, in a rural habitat"

Figure 5: Earth at night / Europe. Credit: NASA

The research investigated the effects of light sensitivity in a variety of fish species, finding that, "European perch, for example, are still able to capture prey at 0.02 lx, the European eel in its glass eel stage avoids light levels of 0.07 lx or less and the common bream is able to detect prey at light levels of 0.005 lx. In tench, melatonin levels dropped to daytime level after a one-hour pulse of 0.03 lx in the middle of the night, which is comparable to a full moon scenario. The pineal organs of the golden rabbfish are able to perceive light intensities as low as 0,1 lx."

This careful research, and much more, should be relevant to policymakers and conservation advocates arguing for stronger interventions on a global scale to reduce light pollution.

Figure 6: Still image from "What is skyglow"- video.
Photo: U.S.Department of Energy.

Figure 7: LA night view from observatory.
Photo: Photographer Andrea Izzotti

Observations made by ground-based observatories periodically study changes in the atmosphere. A more immediate problem with astronomic observations than with climate change is the increase in light pollution (and radio interference). The composition of the atmosphere appears as a light-reflecting factor. The more light scatters in the atmosphere, the more it disturbs atmospheric studies.

Focusing on the economic benefits of urban structures has alienated us from the factors that make us feel at home in the city, including those that allow us to enjoy the nocturnal space and a pleasant sense of community. We need to make the nocturnal spaces of our cities more human.

The biggest single element of a city is its buildings, and their sheer mass has a major impact on the overall nocturnal harmony of the city. Using light correctly is pivotal. Light should serve to make buildings, philosophically, great "whispering" spatial elements that capture our attention. The "whispering light" phenomenon can be achieved with the right mode of illuminating a building, one that's simultaneously protective of biodiversity. This should be part of a city's holistic Darkness Design philosophy, where light aimed directly at the sky against building facades is eliminated, instead allowing a reflection from the facade to appear. In practice, this means that traditional floodlighting is the wrong way to illuminate.

The kind of facade light that protects biodiversity, saves vast amounts of energy, and stimulates the mind actually comes from the building's own interior lighting, professionally implemented. Correctly programmed preset values for interior illumination achieve a harmonious nocturnal illumination. This eliminates light pollution emanating from a direct and powerful wide angle and saves birds' lives by eliminating the intrinsically attractive, excessively strong light intensities in windows, which cause birds to crash straight into the panes and even into each other at distances of 0-30 meters from the windows.

Easily added programming systems can be used to pre-set lighting values in existing interior lighting, which eliminates reflections of the landscape that fool birds into thinking glass is terrain they can fly into. Selecting the illumination values for each building in turn affects the interior architecture of the building. When properly illuminated, walls, furniture, plants, etc., become visible through the glass, letting the birds see the room for what it is.

The gentle distributed light that reflects from the surfaces of roads and streets onto the exterior surfaces of the buildings adds an additional element to the "whispering light" phenomenon.

Floodlights for facade lighting

Illuminating large facades takes a lot of light. The large amount of ambient city light must be taken into account in floodlight selection for individual buildings. In the picture below, the façade lighting of the high skyscraper has been implemented with almost a hundred high-power floodlights. The effect of light pollution is shown in the figure on the right. Huge amounts of direct and reflected light from façade surfaces (windows and solid material) illuminate space for no purpose.

Figure 8 left, Figure 9 right: Left: Light pollution produced by skyscraper façade lighting in Times Square, NYC. Photo: Julle Oksanen. Right: "Times Square and Broadway at night from the Empire State Building," by mattk1979 is licensed under CC BY-SA 2.0

Heuristic cost calculation for the skyscraper: Power / 60units x 1000W/unit x 4 facades/building= 240 kW / Annual use 4000 hours / Electricity cost 9.43cent/kWh (0.1€/kWh) / Annual electricity cost €96,000+ approximated annual maintenance cost (lamp + work) 15000€. **Total annual costs €111,000.**

Total annual costs if LED lighting fixtures are used: €96,000 (with approx. same lumen output as using metal halide lamps).

Figure 10: Light pollution from New York City skyline, USA. Photo by Chris-Håvard Berge on Unsplash

Interior lighting as facade lighting

Unlike large, bright, façade-lighting solutions, adjustable interior lighting solutions create a cozy and inviting atmosphere in the urban area. Compared to floodlight installations, the brings great energy savings: according to heuristic calculations (see Nocturnal Master Plan for Cities), this can save up to 96%, with no pointless illumination of the sky. Using this alternative, urban buildings do not compete for attention but create a holistic harmony with the urban structure and space. In some special cases, facades of important buildings may require separate floodlighting, but these buildings should be considered as part of the city's overall Strategy Master Plan design (see: Nocturnal Master Plan for Cities in Chapter 3). It would be desirable for such light-polluting solutions to be selected by those responsible for the city's architectural look, with the involvement of a professional darkness designer.

Figures 11 left, Figure 12 right: Left: Quiet architectural "whispering lighting" in stylish New York City. Photo by Mauricio Chavez on Unsplash. Right: The Telenor Building, awarded the Norsk Lyspris Prize in 2003. Metaphor: House as a lighting unit, interior lighting as both façade lighting and plaza outdoor lighting. Photo: Jan Drablos

Technical lighting models for nocturnal architecture

A biodiversity problem

One of our serious biodiversity issues involves bird collisions with windows. Of about 50 billion birds, as many as one billion die each year from striking windows, in the United States alone[8]. If this problem is to be solved, we need to look at what kind of interior lighting is best for nocturnal architecture and biodiversity. Daniel Klem, Jr., a world-renowned biology professor, has studied this issue for 47 years, including the effect of lighting on bird behavior. His master work effectively and concisely opens up numerous problems related to this research and helped me create the Bird Graph presented later.

---

8   *Solid Air*, graphic p. 35

Comments from bird-collision expert Professor Daniel Klem, Jr.

"First, birds certainly do sleep at night, most species, most of the time. However, many bird species migrate at night (called nocturnal migrants), typically beginning their nocturnal travels about 18:00, mostly traveling at altitudes below 700 to 800 meters, descending to the ground at about 02:00 to rest and feed before continuing on to their destination. For both the European part of the world and North America, there are about 200 species of some five billion individual birds that migrate to tropical areas to spend their non-breeding period. Most of these are nocturnal migrants, exceptions being birds of prey, storks, and modest numbers of others.

"To escape air disturbance, some birds will fly to 3,000 to 7,000 meters, with the bar-headed goose needing to fly at over 9,000 meters to pass over the Himalayas. The record bird altitude is still the Ruppell's vulture that was sucked into a jet engine at 12,000 meters over the Ivory Coast in Africa. All this is to say that, with clear nights and overall good weather without impeding cloud cover, birds typically fly high enough so as not to be influenced (attracted) by urban (city) lights, even from the highest skyscrapers.

"However, when the weather is not clear and cloud cover is low, nocturnal migrants are forced to fly under the cloud cover, and then they come under the influence of high-rise buildings projecting lights into the sky. Even then, the birds are for sure attracted to this lighting, *but they do not collide with the building as a result*. They swirl around the building, moving in and out of the lights, seemingly reluctant to leave, and more likely to strike one another than to hit the building. In this confused state they become exhausted, flutter to the ground, and are now within the canyons of concrete and glass at street level. These birds are now seeking shelter and can be influenced by urban lighting. They are moving about as early as 03:00, and vegetation behind clear glass can attract them. When they try to reach this vegetation, they strike the glass and are often killed or mortally injured.

"As dawn arrives and natural lighting intensifies, the birds in the city are now subjected to the deceptive reflections of vegetation and sky in sheet glass. Remember that even a perfectly clear pane of glass will act like a mirror when covering a dark interior space. Tinted panes are even more mirror-like. Clear panes as railings, corridors (linkways) between buildings, those making up walls around atria, and noise barriers along roadways are invisible clear barriers to birds trying to reach vegetation, sky, or otherwise unobstructed space behind these clear panes. For sure, birds behave as if clear and reflective glass are invisible to them. For sure, under unusual conditions, lights can enhance the danger for birds by increasing their density in the immediate vicinity (within 30 meters) of the glass surface, where they are more likely to be deceived. Light in these ways constitutes increased risk under these relatively unusual circumstances. *Light, however, is not the principal cause of bird-window collisions.*

"Consequently, lights-out programs and policies in North American cities I support, but it is sheet glass and not lights that is killing billions of birds worldwide, with current estimates of well over a billion birds annually in the US alone. One exception that you (the author) are interested in addressing is making lights less of an attractant, especially during the early morning hours, before the increasing light intensity of sunrise, when migrant birds are in city areas. *In the case of this interior lighting that highlights vegetation behind windows, your (the author's) idea to dim this lighting so that vegetation behind sheet glass is hidden or reduced and thus birds are not attracted to it will save many bird lives."*

Figure 13: Daniel Klem Jr. is an American ornithologist known for his pioneering research into the mortality of birds due to glass windows. He is Sarkis Acopian Professor of Ornithology and Conservation Biology at Muhlenberg College, Allentown, Pennsylvania, USA. In his 1990 papers "Bird injuries, cause of death, and recuperation from collisions with windows" and "Collisions between birds and windows: mortality and prevention," he calculated that between 100 million and 1 billion birds are killed, annually, in the United States alone, by flying into windows. His research has influenced the design of buildings, not least at the Niagara Falls State Park Observation Tower, on which he was a design consultant. He holds several US patents relating to window design. Photo: P-G. Saenger.

Legislative measures to promote bird-safe buildings

According to Klem, "An obvious prediction, as awareness of the problem and growing solutions continue to emerge, is that legislation to protect birds from sheet glass will grow accordingly until it is required everywhere."

Walker Glass collected information on such measures, with thanks to Daniel Klem and Micheal Mesure of FLAP Canada, in an article called "A Brief

History of Bird Deterrence Legislation — Three Levels of Bird Safe Glass Legislation (Federal level, Regional Level, Municipal Level) – Projections for the Future."

"Billions of birds are killed flying into sheet glass installed in buildings as clear and reflective windows. No reasonable person condones these unintended and unwanted tragedies caused by one of the most useful and aesthetically attractive building materials. There are solutions to make windows safe for birds and humans. History has revealed that the use of the legal system is a far more powerful means of stimulating action to protect birds from windows than relying on the voluntary efforts of the many constituencies involved in this important conservation issue.

"International treaties, provincial and state regional laws, county, city, and other municipal ordinances and zoning regulations directly and indirectly address preventing bird fatalities resulting from window strikes.

The most prominent international bird protection legal agreements relevant to avian mortality at windows are the Migratory Birds Convention Act (MBCA), Species at Risk Act (SARA) for Canada; their equivalents the Migratory Bird Treaty Act (MBTA) and Endangered Species Act (ESA) in the United States, and the Birds Directive of the European Commission in the European Union (EU). For North America regionally, protecting birds from windows is justifiably authorized under the Endangered Species Act (ESA) as applied in the province of Ontario; the Ontario Environmental Protection Act (EPA); and the B3 Program (Building, Benchmark, and Beyond) in the state of Minnesota."

Effective methods to prevent bird collisions

In a study published in 2020, "Effective and Attractive Ways To Make Building Windows Safe for Birds," Bryan Lenz, Ph.D., explains how to solve the bird-glass collision issue. Lenz is manager of the glass collisions program and director of the Bird City Network for the American Bird Conservancy.

"The good news is that we know how to address this crisis: Add patterns to glass that make it visible to birds, and you will prevent collisions. All buildings are a collision threat. This means that making a real dent in collision numbers requires fixing homes, low-rise, and high-rise buildings."

Lenz's research reveals surprising bird mortality in collisions with buildings of different sizes.

Residences 1-3 floors: 122.9 million buildings / 2.1 birds per building annually / 258.09 million dead birds annually.
Low-rise buildings 4-11 floors: 15.1 million buildings / 21.7 birds per building annually / 327.67 million dead birds annually.
High-rise buildings 12 + floors: 0.5 million buildings / 24.3 birds per building annually / 12.15 million dead birds annually
Total: 599 million birds per year.

Figure 14: Graphic by American Bird Conservancy

Retrofit solutions can help fix deadly windows in residential buildings (houses and other buildings of up to three floors). The managers of some low-rise buildings (4- to 11-floor buildings) are also stepping up to save birds. This is an important group, given the high mortality and large number of buildings of this size.

In his report, Lenz presents five large retrofits that have made a difference, four of which are presented here.

Northwestern University

Figure 15: Francis Searle Hall, Northwestern University. Photos by American Bird Conservancy

The University has applied bird-friendly window film retrofits to Francis Searle Hall and the Kellogg School of Management. Window film installed in fall 2017 has reduced collisions by 95 percent.

Cleveland State University (CSU)

Figure 16: Cleveland Marshall College of Law Library, Cleveland State University. Photos courtesy of Cleveland State University.

The Cleveland-Marshall College of Law Library has glass walls that reflect trees. The retrofit window films, which put 3/8-inch white dots spaced 2 inches apart on the glass, reduced fall 2019 collisions to zero.

Jacob Javits Convention Center (or Javits Center)

Figure 17: Javits Center. Photos by Ajay Suresh (left) and Susan Elbin (right).

With its exterior consisting almost entirely of reflecting glass, it is easy to see why this was once known as one of the city's top bird-killing buildings before a renovation was done that replaced the building's windows with energy-efficient glass. Because this project changed the glass rather than modifying it, it is technically a renovation and not a retrofit, but this is a terrific way to ensure bird-friendly windows for life. The new glass is patterned using frit (a partial layer of ceramic on the surface of the glass) that is imperceptible at a distance but alerts birds to the presence of glass early enough for them to change their flight path and avoid collisions. The renovation not only replaced the existing glass but also added new glass—and it reduced collisions by 95 percent while saving energy, and therefore money, all at the same time!

University of Chicago

"Not every solution for a large building has to involve window films or changing the glass," writes Brian Lenz. "There are many other ways to fix deadly windows. One popular, effective, easy-to-install, and cost-effective method is called Acopian Bird Savers, what's generally known as a Zen wind curtain. This option involves hanging paracord on the outside of windows to form vertical lines that warn birds there is an obstacle to be avoided. (The paracord can also be fastened at the bottom so that it does not move in the breeze). The University of Chicago employed this solution on one of its problem buildings, quickly adding a simple pattern that effectively reduced collisions."

Figure 18: William Eckhardt Research Center, University of Chicago. Photos by Acopian Bird Savers

Canada's example: a pioneer in "bird-friendly building design"

An extremely comprehensive Canadian technical committee's "Bird-friendly building design" standard, CSA A4-460:19, addresses bird-friendly construction design for both existing and new buildings. The aim of the standard is to reduce the number of birds colliding with buildings. It presents requirements for bird-friendly design in the glazing of buildings, structures to be integrated into these, and for general construction and use of building plots, from the perspective of eliminating bird collisions. The numerous requirements relating to Standard CSA A460:19 are precisely and clearly delineated. It does not include other standards pertaining to building construction such as energy efficiency, residents' convenience or the safety of glazing.

The negative effects of lighting are addressed in the standard at 3. Bird collision mitigation strategies, and 3.6 Lighting, which covers both exterior and interior lighting. Annex A (Informative) of the standard, "Bird-friendly building design – Overview and background rationales for requirements"; A6 Light Pollution specifies the effects of light pollution at a generally comprehensible level, considering, among other things, the effects of lighting on the behavior of migrating birds while migrating and also how little is known about the interest birds have in light and how this affects a bird's orientation mechanisms.

> **The Canadian national standard: CSA A460: 19**
> 3.Bird collision mitigation strategies
> 3.6 Lighting
> 3.6.1 **Exterior lighting: All exterior building lighting shall be dark sky compliant.**
> 3.6.2 **Interior lighting: A building's interior lighting should be reduced** after business hours in non-residential buildings and from sunset to sunrise in all cases. Whenever possible, **task lighting** rather than building lighting should be used during these times.

Lighting to avoid mirroring

The most important feature of all window glass is transparency. Yet all windows, even perfectly clear panes, will act like a mirror, reflecting the facing habitat and sky, when outside lighting is more intense than the darker indoor space the window covers. Outside illumination levels normally vary in the range of 3000lx–50000lx, and indoor office lighting illumination levels vary from 100lx to 500lx. The contrast is huge, and interior space looks dark from outside and the exterior super bright from inside. The window will act like a mirror for birds—like "solid air."

Light produced inside the building compensates for the mirror reflections that attract birds if the difference in light level is small (that is, during twilight but not in the daytime). Mirror reflections can be prevented by illuminating interior surfaces (walls, tables, ceilings, and so on) and not the window glass itself. Therefore, light should be produced "deep" inside the building (with reflections in many directions). Ideally, the illumination levels of this "compensation light" will correspond to the changing times of twilight through the year as well as technical features of the building lighting system. Illuminating the window glass itself, for example with wall-washer type light fixtures near the window, would be a bad solution—totally useless—and would only cause light pollution outside the building while attracting birds to fly toward the bright light.

Figure 19 left, Figure 20: When the interior surfaces of a building are illuminated, light is reflected from each surface several times and hits new surfaces repeatedly. Light does not "disappear," but eventually turns into heat; some of it escapes from the windows. Left: Interior lighting compensates for the "mirror effect" at twilight. Right: glass surfaces produce the mirror effect in bright daylight. Photos: Museum of Contemporary Art/Kiasma/ Senaattikiinteistöt

The "Light, Glass ('mirrors') and Bird Behavior graph

Among the challenges Darkness Design aims to meet is the need to protect birds, in part by considering how buildings are positioned on our revolving globe, with its constantly changing light.

Good interior lighting design aims to produce a "whispering light" effect, an architectonically demanding concept. Designers need to have some understanding of bird behaviors in response to various strengths of interior lighting at different stages of twilight. Everyone engaged in such design has a responsibility to protect birds as part of the biodiverse environment by, as much as possible, eliminating mirror reflections from window glass (which occur because of a low interior lighting level) or producing excessively powerful interior lighting that actually draws birds to the glass.

Variations in the season and time of day cause constant changes in natural light. These changes can be described this way: night—astronomical dawn—astronomical twilight—nautical dawn—nautical twilight—civil dawn—civil twilight—sunrise—noon—sunset—civil twilight—civil dusk—nautical twilight—nautical dusk—astronomical twilight—astronomical dusk—night. The effects of these varying levels of dimness in relation to the "whispering light" emitted from windows can be controlled by new lighting control technologies (discussed later in this book), both in buildings equipped with old lighting technology and in new buildings.

Even though today's control systems make implementing the correct lighting values relatively easy and inexpensive, the people responsible for this need motivation, mental commitment, and a true appreciation of the urgency of this task, including awareness of the fact that the total number of birds in the world has diminished by seventy percent in the last fifty years.

Many of the international studies I have read on bird collisions with windows suggest that the role lighting might play in preventing this is unclear, in part because daytime lighting cannot be used to counteract mirror reflections, in which the flying bird sees in the windows an interesting landscape into which it wants to fly. I have found in the research no pragmatic compensatory values for interior lighting, only comments about indoor lights that attract birds at night and various proposals for lighting to compensate for this.

However, to help designers prevent bird collisions as much as possible, there are easily implemented interior lighting values that correspond to the respective degrees of dimness in exterior light.

A splendid joint thesis subject for young students of engineering, biology and architecture would be to investigate the effects on bird collisions and resulting deaths of implementing lighting solutions at different degrees of dimness (with and without retrofitted windows). Collaboration among rising stars in different disciplines would help to remove barriers noted earlier (Human Resistance to Change) and serve as an example to all. This could be a basis for new and groundbreaking research work.

24 hours in the life of birds

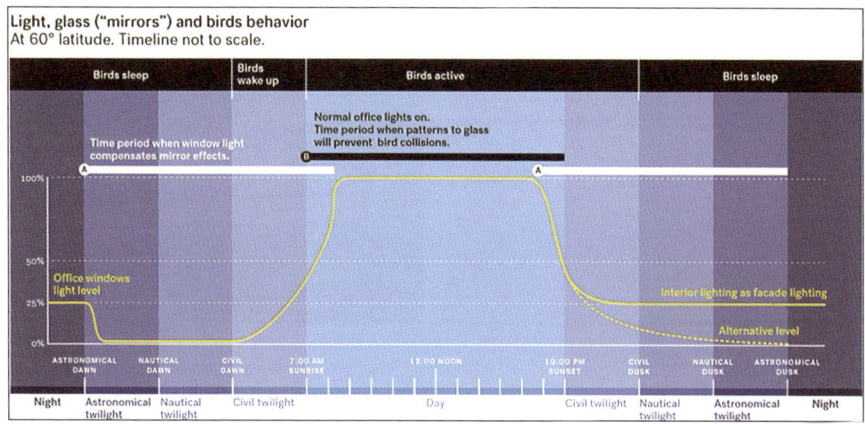

Figure 21: "Light, Glass ('mirrors') and Bird Behavior" graph. A biodiversity-friendly and energy-saving model for nocturnal architecture and all building types. Photo: Author and Safa Hovinen

The graphic includes all the elements that affect the lighting values of interior lighting in nocturnal periods. Indoor lighting is the right solution for facade lighting in public buildings because it minimizes the effects of light pollution on the sky. The yellow line on the graph represents the ideal use of interior lighting. It takes into account the average behavior of various bird species (mainly passerine) at different times of the day, the characteristics of outdoor lighting at different times of the day and their effects on the ideal amount of light produced. Lighting can compensate for mirror effects when sufficient interior lighting illumination values are used to reveal an interior architecture that birds will not be interested in. It is possible that indoor lighting that's too bright relative to the intensity of outdoor lighting may seem too inviting, and birds may steer their flight towards dazzling bright windows, which creates a clear risk of collision.

Birds sleep–Birds wake up

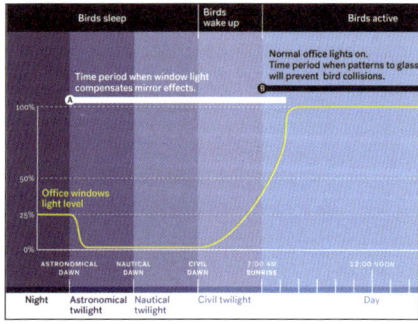 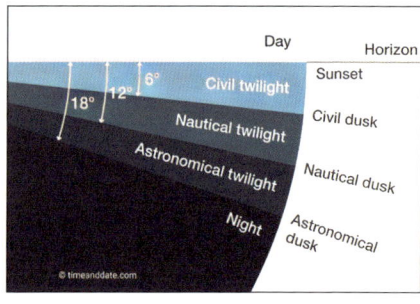

Figures 22 Left Figure 23 Right: Morning twilight: Astronomical, nautical, and civil stages at dawn. Photo: Wikipedia

As the sun rises towards the horizon, light from the sky awakens birds (some as early as during astronomical twilight). Birds are active before sunrise, and the risk of a collision with windows increases considerably if the building's interior lighting is not on and the environment is of interest to birds. Birds fly here and there looking for food, interacting with other birds, or just because they are happy and full of energy. Before sunrise, the outdoor lighting intensity increases relatively quickly and strongly, to a value where windows turn into mirrors for birds. Lighting intensities vary by location, season, and weather. Just before sunrise, illumination values easily reach values of hundreds of lux. To compensate for the mirror effect, it's important to increase interior lighting illumination values at this point.

After sunrise, the illumination values of outdoor light reach thousands of lux, making it impossible to compensate for the mirror effect with indoor lighting. In offices, normal illuminance value is typically around 300lx. A heuristic calculation of the proportions between indoor and outdoor brightness values is easy. Outdoor lighting luminance (we see this as brightness) is between 300 cd/m2 and 8300 cd/m2, and the indoor maximum, for example in an office, is 50 cd/m2. During daytime, the only way to save birds from collisions and death is to put patterns on the window glass.

During the astronomical dawn and civil dawn, compensation values for indoor lighting should be set at minimum levels, so that indoor lighting does not become too attractive to birds. Illumination values could be just

15-20 lx. According to Dr. Raimo Hissa's article "Birds vision,"[9] a bird of prey can see its target 10 times better than humans. Birds' eyes are also much more sensitive to the color spectrum than those of humans. Birds can see ultraviolet radiation, for example, because they have an "oil drop" in their eyes so blue that ultraviolet radiation does not scatter on the retina, but sharp light beams continue straight to the retina. We humans don't have this "oil drop," which is why we don't see ultraviolet light at all, and blue light scatters into the human retina.

The correct compensation values for interior lighting need to be tested on a case-by-case basis, as there are many variables (building location, window location, interior structures, and materials that reflect light onto other structures such as walls, tables, ceilings, interior lighting structures, lighting types, lighting control systems, etc.). We'll offer some practical solutions in the section "Technical lighting models for nocturnal architecture."

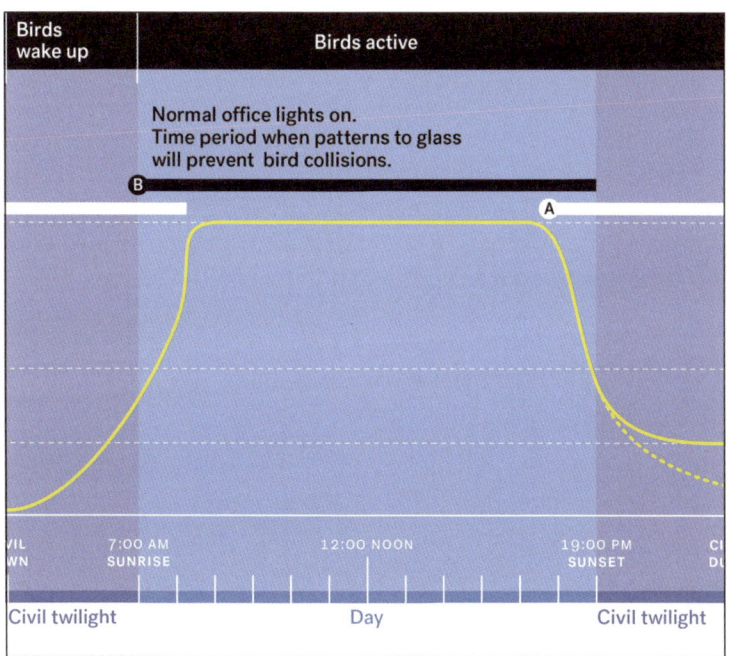

Figure 24: Birds' active period and light

---

9   Birdlife.fi, page 160

Figure 25: National Wildlife Federation. Photo: Deborah Allen

During the day, even if the "normal" office lighting is on 100%, windows act as mirrors. The ratio of daylight to electric lighting is almost incomprehensibly high. On a sunny day, light intensity values outdoors can be up to 300 times greater than office light values in full light. During daytime, when birds are active, the fascinating landscape from which the bird flies towards the window is reflected from the window onto the retinas of the bird's eyes.

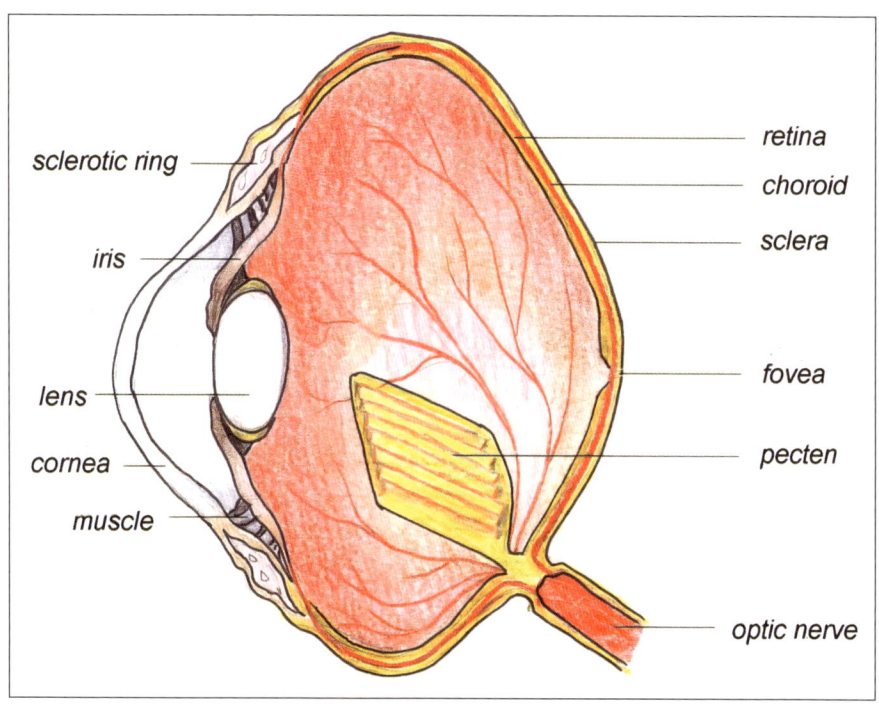

Figure 26: From Wikimedia Commons, the free media repository

Optic nerves on the retina receive different kinds of electrical impulses produced by electromagnetic radiation from different colors of light (objects such as leaves are green, tree trunks brown, sky blue, etc.) reflected from the window. Electrical impulses are directed along the optic nerves to the bird's brain, which perceives the mirror image as an inviting landscape, possibly offering food and companions. The bird flies towards the window, completely unaware that the landscape it sees is in fact hard, deadly glass. In this daytime period, only patterns on the glass will prevent collisions.

Birds sleep

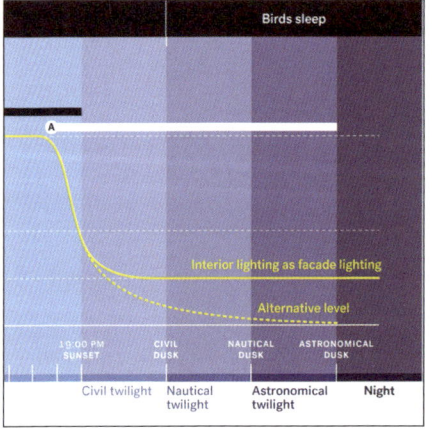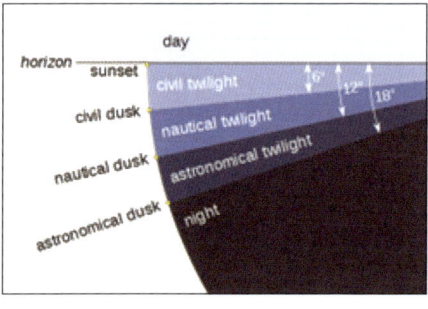

Figures: Left 27, Right 28. Evening twilight: civil, nautical, and astronomical stages at dusk. Photo: Twilight

According to ornithologist Daniel Klem, Jr, birds may well be flying during the astronomical twilight period just before the darkness of night. As shown in the graph, workers might leave their offices starting at 4 p.m., randomly turning off the office lights unless they are governed by an automatic lighting control system. Studies show that automatic controls do this more reliably than humans. In most cases, lighting in the corridors, which produces lighting intensities of about 100lx, is left on. This level of corridor illumination creates an adequate "collision protection level" for post-sunset twilight (civil twilight, nautical twilight, and astronomical twilight). In modern offices, the doors are often made of glass. If the doors are of solid material and we are relying on corridor light, individual office doors must be kept open so that corridor light can be reflected from the various structures of the offices, such as walls, tables, and ceilings. From the perspective of birds, this prevents mirror reflections occurring in office windows and thus prevents fatal collisions with windows.

An outlet-mounted light with an adjustable astronomical clock is an ideal solution, because little light is needed to achieve these results. A more cost-effective option is to keep workstation lights on, either in a controlled manner or at all times (LED table lights).

A yellow dashed line in the graph suggests an alternative level. When the office lighting acts as facade lighting for a building, a 25% level of office lighting is ideal to create a peaceful-looking building. In most cases, urban street lighting also creates an adequate and peaceful addition to facade lighting. However, it is unnecessary to keep this 25% level of facade lighting on when people are sleeping, and with light control systems, office lights can be switched off and only the corridor lights remain on.

Technical lighting examples for nocturnal architecture

Let's turn now to some pragmatic methods that meet biodiversity-friendly requirements with skillfully implemented nocturnal architectural design (see "Light, Glass and Bird Behavior" graph), while meeting the goals of the internationally acclaimed Canadian national standard (CSA A460: 19 "Bird-friendly building design"). The Canadian standard covers ways to mitigate harm to birds involving office buildings but should apply to removing the mirror effect on windows of all types of buildings and to adapting buildings to nocturnal architecture.

New lighting control systems are suitable for both new and old buildings. International systems suitable for renewing lighting units and controls include so-called addressable control systems, such as the Digital Addressable Lighting Interface (DALI) or KNX systems. Wireless property lighting control systems can also be used, for example Zigbee. Lighting system upgrades can be implemented with wireless control systems by replacing some existing fluorescent lamps with LED lamps that include a ZIP receiver. Internationally known wireless control systems include, for example, Zigbee or Bluetooth.

It is noteworthy that just a few lamps per building floor are needed for this task when they are placed "deep" inside each floor, so that the reflected light flux entering far from the interior surfaces (walls, ceiling, and other elements) is enough to compensate for the reflections from windows that birds see. After checking the minimum brightness level needed to compensate (visually, and possibly also by measurement, which will help engineers strategize for selecting lighting levels and lumen production/dimming levels), the selected lighting units can be preset by computer or phone to the correct values—for example, 10% of max during the morning twilight times in the graph—to cover the time from sunset to sunrise.

A high-rise example: Old fluorescent lighting / Seagram Building, New York City

Figure 30: "Seagram Building," by Tom Ravenscroft, is licensed under CC BY 2.0.

The lighting design principles of the Seagram Building, completed in 1957, were unique and it was an icon of its time. According to the book *The Structure of Light, Richard Kelly and the Illumination of Modern Architecture*, the lighting was designed by Kelly in collaboration with the architects Mies Van der Rohe and Philip Johnson. During the day, the panel ceiling's primary lamp circuit produced roughly 915 lx, ostensibly to counter glare from the outside. In the evening, the secondary circuit along the perimeter used separate lamps running at one-quarter power to produce light levels near 215 lx. These lights were intended to be seen from the sidewalk, and thus allowed for the illumination of the tower as a singular unit.

Today this lighting intensity corresponds to the lighting value of modern offices during the day and is architecturally too high and dominant in the evening. With its brightness, it attracts birds to collide with the windows.

1. Possible 2000s modernized panel lighting solution:
- During the day, general lighting level from 915 lx down to 200-300 lx.
- During dark period, lighting level from 215 lx down to 25 lx.
- During migration periods, the building must be left in complete darkness (dark compliant), as with all skyscrapers.

2. Benefits:
- Existing solution still useful.
- Energy savings 80% on average.
- Light-pollution-free maximization.
- Bird-collision-free maximization.
- Architecturally interesting "whispering light" solution.

3. Challenges and tasks for the lighting collaboration team

Architect: Visual responsibility.

Darkness (formerly lighting) designer: Arguments for final "whispering lighting" levels (elimination of mirror reflections to save birds, and architects' acceptance); producing the final lighting design report.

Electrical engineer: New lighting control system design (DALI, Zigbee, replacing existing fluorescent lamps with LED lamps that include a ZIP receiver, etc.).

Biologist (e.g. person from Birdlife International): Delineating bird migration periods and routes, importance of preventing mirror effect in windows.

Owner: Calculation of costs and payback period; approval of changes.

Glass consultant: Providing information on glass material and possible tests, together with a lighting designer, an electrical engineer and an architect (blinds, earning LEED points, patterns on windows).

City planning office official: providing the City context.

**Residential building example (1-3 floors):
Modern LED lighting solution**

Figure 31: Cleveland Marshall College of Law Library, Cleveland State University. Photos courtesy of Cleveland State university.

1. Possible bird-friendly solution
- During the day: retrofit window film reduces bird collisions.
- During dark period: "alternative level" (see yellow line on figure) value on. Testing is important to check mirror-effect elimination to avoid bird collisions (sufficient contrast to reveal the white dots of the retrofit film), but mainly achieving architectural "whispering light" level, if building is selected as part of the City Master Plan design.

2. Benefits:
- Existing solution still useful.
- Bird-collision-free maximization achieved by allowing birds to see the white dots.
- Architecturally interesting "whispering light" solution.

3. Challenges and tasks for the lighting collaboration team:
Architect: Visual responsibility.
Darkness (formerly lighting) designer: Producing arguments for final lighting levels (elimination of mirror reflections to save birds, and architect's acceptance), producing the final lighting design report.

Electrical engineer: Possibly the new lighting controlling system design. (e.g. DALI, Zigbee, replacing existing fluorescent lamps with LED lamps that include a ZIP receiver, etc.).

Biologist (for example, person from Birdlife International): Delineating the importance of preventing mirror effect in windows.

Owner: Calculation of costs and payback time.

Glass consultant: Providing information on glass material and tests together with a darkness (lighting) designer, an electrical engineer and an architect.

City planning office official: Setting the context for the building in the City Master Plan.

Orchestrated Darkness Design in a high-rise: Telenor Building, Oslo

Figure 33: Telenor Building in Oslo, Norway. The total area of the glass facades is about 15,000 square metres. Photo: Telenor.

Creating a successful and enjoyable nocturnal space requires the designer to orchestrate the darkness instead of spreading excessive light. This project demonstrates how successful darkness can be created.

An orchestrated concept of darkness (Darkness designers Vesa Honkonen & Julle Oksanen)

This office building is located near a beautiful fjord and in a natural setting. As darkness designers, we see Norwegians as people of nature. When we started to create this concept, we had in our minds an image of Norwegians wandering their mountains and fjords wearing their characteristic Norwegian pullovers and rucksacks.

However, we kept in mind that Norway is also very wealthy, with a lot of oil and great educational opportunities. It is a modern nation, of which this Telenor Building is a good example. This is somehow an odd combination for invoking the lighting/Darkness Design concept. Seven thousand employees enter this modern, high-technology building every day. They approach the main entrances through a huge plaza, approximately the size of four full-size soccer fields.

Eight main entrances are located on the oval glass edges of the two building blocks. Passing through these, staff arrive at an indoor entrance made of glass, steel, and stone. After this, they proceed to their own office desks using stairs or elevators.

The next figure shows an architectural section of the office entrance hall block. Made mainly of glass, it is located between the office blocks and includes a cafe.

Heuristic metaphors

We started the "metaphor" process in this design by thinking of what Norway is. Norway has a lot of darkness and beautiful and wild nature, including the stars and Northern Lights. We wanted to bring the darkness from the fjord to the site, letting it float silently through the building, returning to freedom through the eight glass entrances. We wanted to use the Telenor buildings themselves as huge lighting fixtures of the plaza. We wanted the glowing facades to form "whispering lights." Our dream was that when darkness engulfs the site and pervades the entrances to the back of the buildings, the glass facades glow modestly, shedding vertical light on the plaza. We wanted no lighting fixtures on the plaza itself. In this new design, indoor lighting becomes outdoor lighting, as light and darkness are paired in life.

Based on that metaphoric poetry, our concept was to create a place for employees where they could have a clear and visible connection to nature. Where a person in the atrium cafe could look up to find the stars and moon in the Nordic sky.

Glass, as pointed out earlier, acts like a mirror. The reflection factor of normal "float" glass is something like 20%. The hall block was full of revolving glass elements set at different angles both vertically and horizontally. We therefore decided to use a functional lighting concept, meaning that we wanted to minimize reflections from the glass surfaces. These entrance halls were actually "light traps" during dark periods. We used large surfaces and placed lighting fixtures in such a way as to avoid or at least minimize reflections.

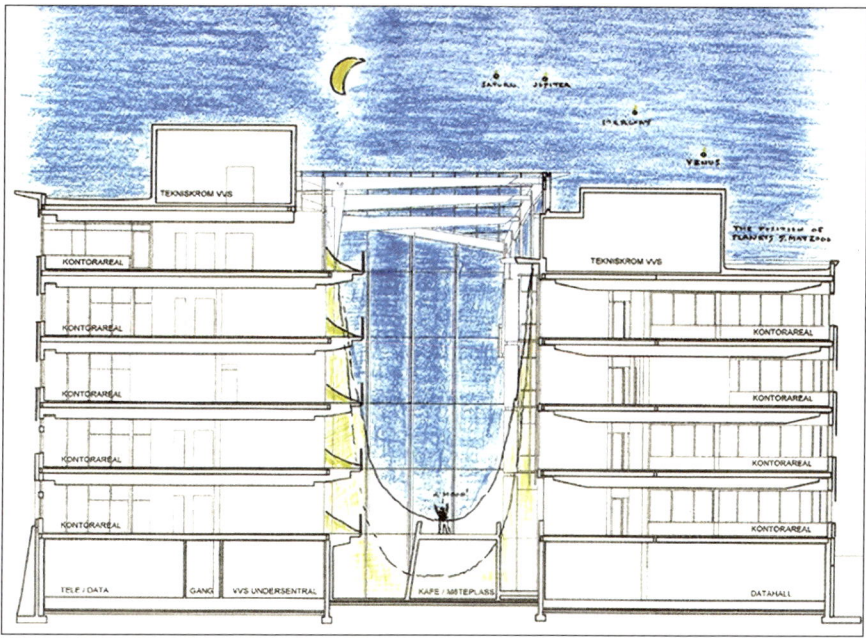

Figure 34: Telenor darkness/lighting design concept. Atrium section. Employees can see a northern sky and stars from the cafe. No glare, no disturbing light distribution surfaces. Photo: Vesa Honkonen & Julle Oksanen Telenor report.

It would have been easy to integrate downward lights on the beams in the roofing area. But we did not want to illuminate only air, which is an invisible material. We also estimated that this common solution would have produced hundreds of reflected light sources.

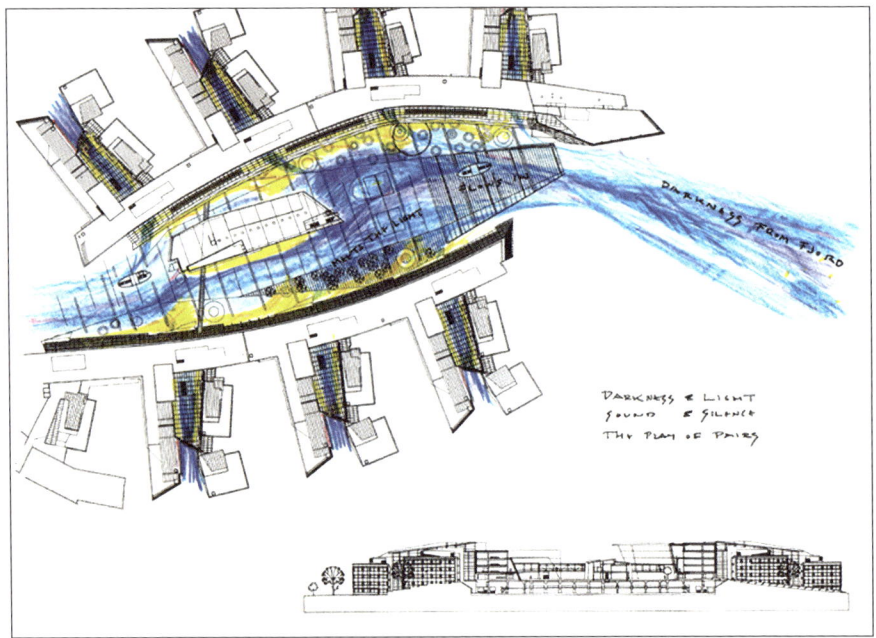

Figure 35: The conceptual philosophy was to bring the darkness from the fjord, letting it flow through the plaza, into the building and out again to freedom. Metaphorically, the building itself is a luminaire. Design: Vesa Honkonen & Julle Oksanen. Photo: Honkonen & Oksanen Telenor report.

- Love & hate
- Sound & silence
- Black & white
- Light & dark

These are the complementary pairings of life. Without these dualities, life would be boring and dull. Shadow is light's best friend, and glare is light's worst enemy. We desperately wanted to apply this philosophy on the plaza. We wanted to bring darkness from the fjord to the Telenor Plaza, carry it through the entrance halls and return it to nature. This is somehow a very Norwegian style. This concept meant that we didn't want any light fixtures on the plaza. The huge glass openings of the building blocks were our large illuminators. The education center, which is located on the plaza, also had glowing walls. This concept meant that we had to be able to influence and use the office lighting effectively.

Orchestrating the Master Plan of Darkness (Darkness designers Vesa Honkonen & Julle Oksanen)

The master plan is an important strategic tool in the Darkness Design process. Darkness designers have to find lighting hardware and software solutions that can implement the concept on a practical level. As this stage, they will need to choose the right lighting units and lamps, do preliminary checks on construction solutions, do lighting calculations, create computer images, etc.

If the right solutions cannot be found, it may be necessary to change the concept, or even create a new one. With accumulated experience in real-world projects, a darkness designer grows in understanding of what can or cannot be implemented. Sometimes a great concept has to stay on the level of dreams. But one should not give up hope. A minor change in design need not mean the irretrievable loss of a great idea, and may still produce a high-quality result.

The master plan phase doesn't yet involve detailed design elements such as working drawings, construction detailing and lighting fixture integration into structural elements. It's a good time to clarify what is in the offer, to avoid misunderstandings later about what is included in the work and what is not.

Some examples of the master plan stage (Details are further fleshed out in the Appendix.)

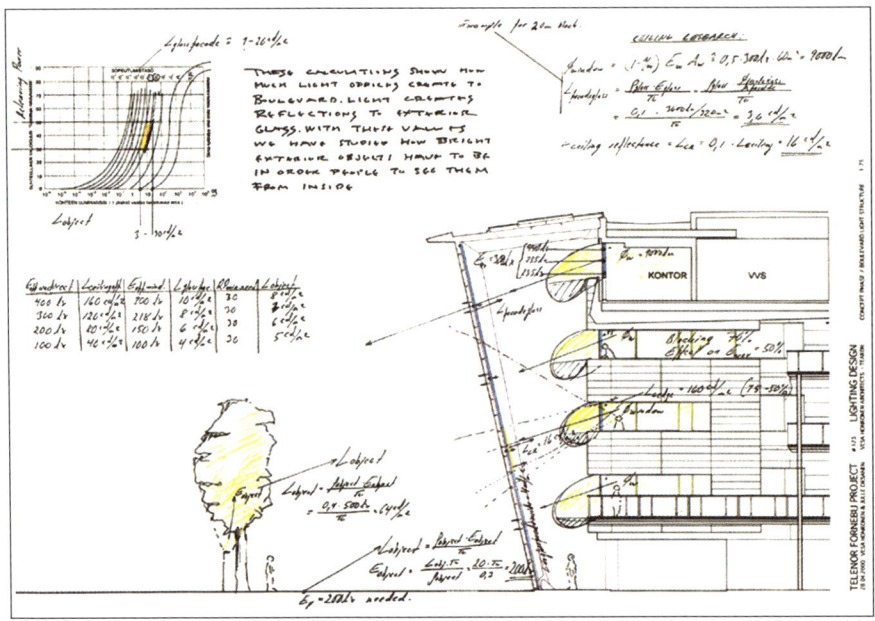

Figure 36: Numerous calculations were needed to achieve the darkness flow inside. Photo: Honkonen & Oksanen report.

Figure 37: Flowing darkness. Photo: Honkonen & Oksanen report. Credit: Jan Drablos

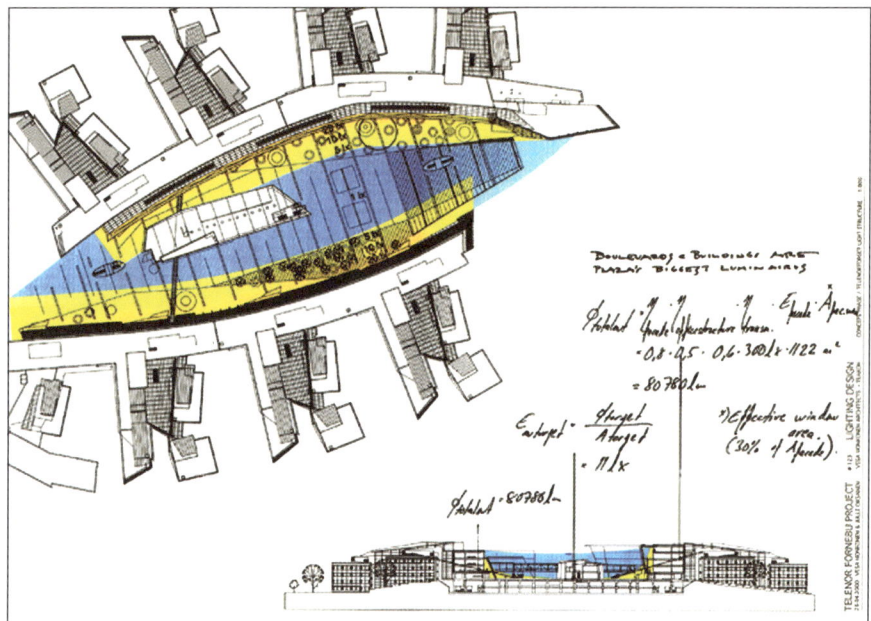

Figure 38: Many simple, basic calculations went into achieving the darkness flow outside so that no lighting fixtures were used in outdoor areas. Photo: Left, Honkonen & Oksanen report

Figure 39: The darkness flow outside to inside. No lighting fixtures were used in outdoor areas. Photo: Jan Drablos.

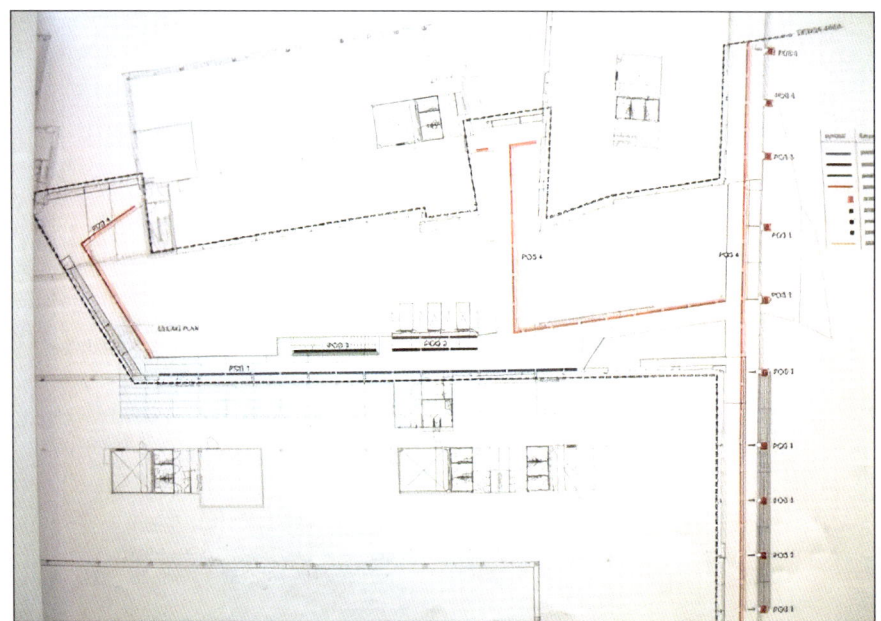

Figure 40: Lighting was carried out with 8 kilometers of Notor luminaires: Direct and indirect versions.

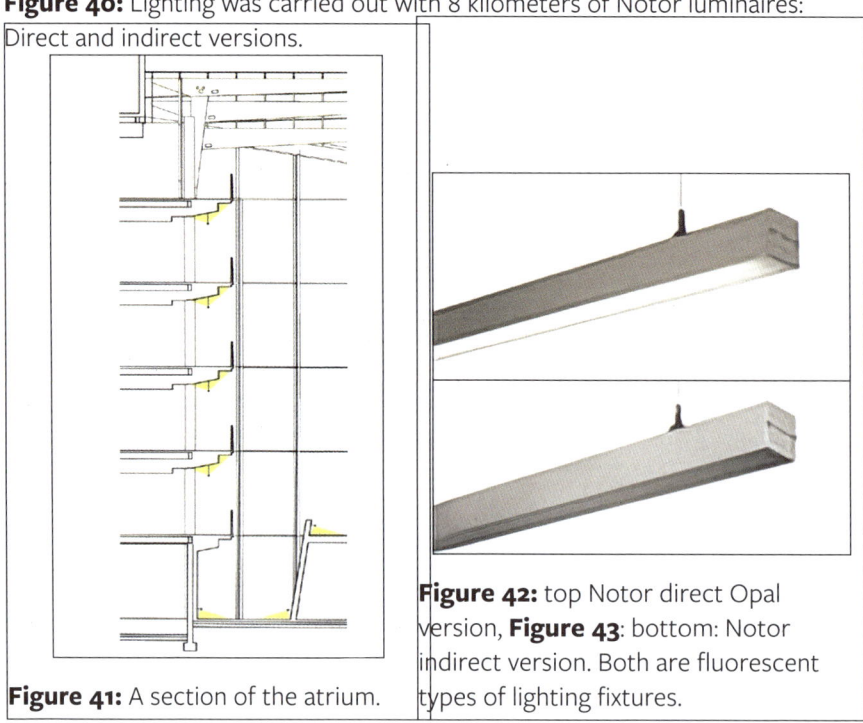

Figure 41: A section of the atrium.

Figure 42: top Notor direct Opal version, **Figure 43**: bottom: Notor indirect version. Both are fluorescent types of lighting fixtures.

Challenges and tasks for the lighting collaboration team

Task: Eliminating reflections from window areas at night by minimizing interior lighting and adjusting the system as needed. (No changes needed in this project.)
Architect: Visual responsibility.
Darkness (formerly lighting) designer: Presenting arguments for final lighting levels (elimination of mirror reflections to save birds and architect's acceptance; window markings + "whispering" lighting levels), producing the final lighting design report.
Electrical engineer: New use of lighting control system, taking into account the Light, Glass ("mirrors") and Bird Behavior chart and the time employees spend in the plaza (the building is used on a 24/7 basis).
Biologist: delineating bird migration periods, importance of preventing a mirror effect.
Owner: Calculation of costs and payback time calculus.
Glass consultant: Providing information on glass material and testing, together with a darkness (lighting) designer and an architect.
City planning office official: Presenting the City context.

Note: The full project design can be found in Appendix 1.

Upwards-Aimed Facade Lighting: conclusions

Carefully programmed and designed interior lighting for buildings protects biodiversity and offers the right way to illuminate building facades during twilight times. This eliminates bird strikes on glass and unnecessary light pollution emanating from the nocturnal architecture of various urban environments.

Other human structures and light pollution

Tall structures and flight obstacle lighting

When powerful floodlights are employed, airborne particles cause strong light phenomena. Lighting beamed into the night sky for informative purposes, such as in the post-9/11 memorial installation "Tribute in Light," is often cited as an example of this, though this sort of thing is most often short-lived. Mostly, this kind of lighting is used merely to offer "something spectacular." If such floodlights are used while birds are migrating, it causes confusion and makes them fly in and out of the light, often provoking collisions, falls and death. Using variously colored lights is another unjustified lighting solution. Inhabitants of some Middle Eastern and Asian countries are already tiring of the expanded use of huge, colored public lighting and bright colored lights aimed at the sky, based on my personal lighting design experiences in these areas.

Figure 44: "9/11 Tribute in Light," by karol.cc, is licensed under CC BY-SA 2.0.

Figure 45: "City Lights," by Girl flyer, is licensed under CC BY 2.0.

Natural rhythms and "whispering" lighting

We need to start repairing our living environment in a more natural and caring direction by removing the light pollution caused by poor design and implementation of electric lighting. For that we need to deepen our understanding of life in general. In his book *The Core, Better Life, Better Performance*, Aki Hintsa, MD, talks about the importance of a regular daily routine. He worked for more than 10 years as a Formula 1 paddock physician, a Finnish Olympic Committee physician, a referral physician in Ethiopia, and at his own specialty clinic in Geneva, assisting international business leaders in their lives.

Dr. Hintsa lived in Ethiopia for long periods and became friends with the world's toughest endurance runner, Haile Gebrselassie, even living in the runner's home as a family member. "Dr. Aki," as Haile called him, made an important observation that guided his professional mission throughout his life: the importance of following a regular circadian rhythm that closely follows the sun's movements from day to day and season to season. Gebrselassie's happy and balanced family rose every morning before sunrise, at about five o'clock. After his daily routines (running and gym workouts, running a business and social contacts, happily and energetically caring for others)

Haile returned home at about 5 p.m. After a family dinner at 6 p.m., the family went to sleep when the sun went down.

Lack of adequate sleep is one of the most significant health risks in modern society. We need eight hours of it daily. Dr. Aki Hintsa's consultations with patients almost invariably focused on getting adequate sleep. The executives of large companies and high-level politicians almost invariably slept less than eight hours, which Aki Hintsa often analyzed as the reason they ended up at his Geneva clinic.

We humans, like all animals, naturally need night and darkness. During normal nocturnal periods, architecture doesn't need to have screaming colored lights and unnecessary illumination of the sky. For example, Niagara Falls, in nocturnal periods after the artificial lighting has been turned off, has soft whispering natural light from the sky, accompanied by mystical mist and the sound of falling water—the natural beauty of life.

Without noticing it, we have lost our soul connection with the world, losing one more piece of darkness after another, over more than a hundred years. We will discuss "whispering" nocturnal Darkness Design more fully in Chapter 3, Paradigm for Design of Nocturnal Spaces.

Figure 46: "Niagara Falls American Falls lit at night," by Nayukim, is licensed under CC BY 2.0.

Figure 47: "Horseshoe Falls, Niagara Falls, Ontario," by Ken Lund, is licensed under CC BY-SA 2.0.

Wind turbines and obstruction lights

For many bird species, individual electric lights provoke changes of direction, attraction, communication problems, reproductive problems, and fatal collisions with each other and with wind farm structures. *Solid Air Invisible Killer*[10] estimates that, "In the United States in 2008, masts and wind farms, as well as power lines invisible at night, caused 6.8 million bird deaths." In Finland, the reference figure corresponds to 38,633 dead birds. Migratory routes for birds should be taken into account seriously when designing tall structures.

Figure 48 Left, Figure 49 right:
Left: "Wind Farm and Milky Way 2," by fireboat895, is licensed under CC BY 2.0. **Right:** "Stop these things"[11].

---
10 Prof. Daniel Klem, Jr., PhD. p. 35
11 https://www.sourcewatch.org/index.php/%22Stop_These_Things%22

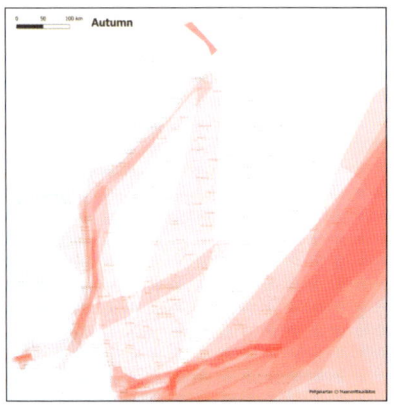

Figures 50 left: Migratory routes in Finland.
Figure 51 right: Migration Challenges and Conservation. Credit: Birdfact.com

Luhanka wind farm, Finland

The lights of the Luhanka wind farm, located in the center of the map (green triangle) appear on the horizon point for migrating birds on an intersecting flight path.

Figure 52: Horizon calculation: hO is a flock of migratory birds flying 800 m high, below the clouds. d1 is the distance at which a flock of migratory birds sees on the horizon, for example, the red obstacle light of a wind farm **d1 = 101 km**.
Credit: Pauliina Oksanen

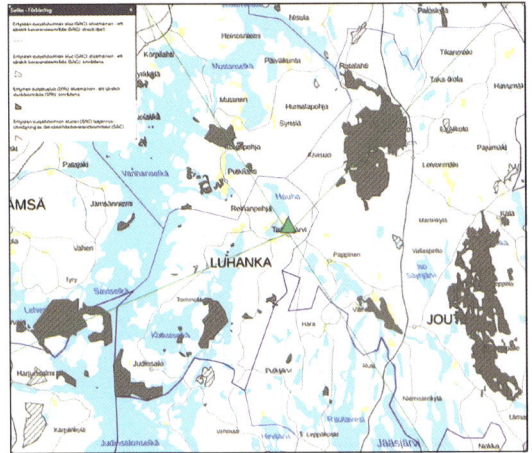 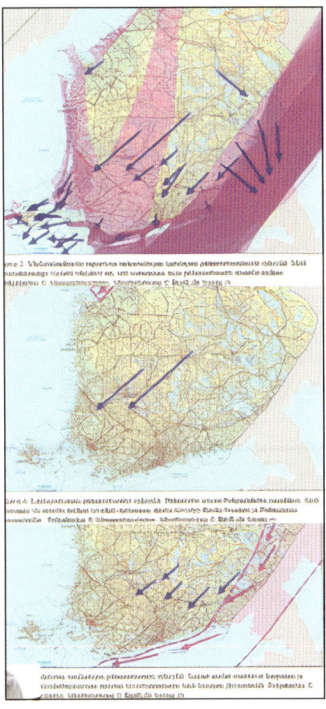

Figure 53 left, Figure 54 right
Left: Luhanka wind farm location in center of map (green triangle). Map size 80km x 80km. Shaded areas on the map are natural preservation areas.
Figure 54: Right: The main migration routes of various bird species. All migration routes intersect above Luhanka.

When flying within a distance of 100 km (an area that doesn't even fit on this map) a flock of migratory birds will see red wind-farm obstacle lights on their night migration and may find them interesting enough to direct their flight towards them, with potentially devastating consequences as birds fly into turbines in the darkness.

Challenges in solving this problem: Research is needed into what makes points and elements of light so attractive to a bird that it will fly toward them, as well as the maximum range for such attraction and how devastating the consequences of these lights are for birds when flying on migratory paths. It might be possible to replace warning lights with a signal connection between aircraft and wind turbines instead.

UPWARDS-AIMED TREE LIGHTING

In the 1990s, tree lighting became a universal way of enlivening cityscapes. This was due in part to the evolution of lighting design as a profession. Light pollution on a large scale was a virtually unconsidered concept among professionals (architects, engineers, and lighting designers) and the general public alike. Trees and parks were rewarding elements to illuminate in urban spaces.

Tree lighting is best considered by means of an example, namely the Turku Aura River Project, in which 100 riverside oaks were first illuminated in 1997-8. We'll look at this project and then consider how it might be done today in a way that values biodiversity.

Aura River tree lighting (as implemented in 1997)

Architect Vesa Honkonen and the author received this major lighting design project in the Finnish city of Turku in 1997, the task being to create a lighting design for a scenic section of the River Aura. For the first phase of the project, we selected 100 large trees along the river, located between bridges, to beautify the riverbank and the cathedral. Upon completion, the project received domestic recognition and became celebrated abroad, mainly in the USA.

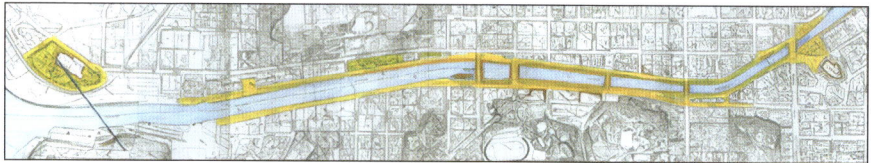

Figure 55: Aura River Project, Phase 1, photo: Aura River Report

After researching numerous lighting options to illuminate the trees in this historic area, we came to the conclusion that among the thousands of existing lighting options, there were none at all that showed appreciation for space. There were large, ugly lighting fixtures, dazzling floodlights, fire-hazardous recessed luminaires, and innumerable other types. Thus we immediately ran into a huge problem on our first major project, and at the same time a huge challenge. The River Aura is the beating heart of Turku and a major source of local pride. Comprehensive expertise was called for.

Figure 56 left, Figure 57 right

Some examples of existing lighting options:
- Ugly, big and glare-y floodlight that also called for a glare shield and an ugly one-meter pole. A hundred such lighting structures in the historic and sensitive center of Turku would have been a disaster, and the project would have been over before it had even begun. It would have destroyed our reputation as designers.
- Flammable underground model. In the late 1990s, the lens of this large, recessed lighting unit easily heated to over 100 degrees. Each tree would have required two underground light sources, resulting in 200 flammable barrels just waiting for dry hay, debris, and dog feces to ignite. Photo: Aura River Report

Testing our unique design method and lighting fixture

As the duly appointed architectural lighting designers on this project, we applied this philosophy:

"Shadow is light's best friend." (A slogan drawn from the author's dissertation "Design Concepts in Architectural Outdoor Lighting Design, Based on Metaphors as a Heuristic Tool.")

Architectural outdoor lighting design, as a process, is at the same time also darkness and shadow design. Darkness fascinates us as human beings, but its design needs exceptional skills, open-mindedness, and courage.

Let's imagine a hypothetical situation in which the lighting designer begins to design a lit environment with total darkness as the starting point. The lighting designer begins to remove, or eliminate, dark layers from the totally black background, removing darkness layers one at the time, until the desired lighting degree on the designed surface (e.g. on a façade or tree) is achieved. The lighting designer proceeds layer by layer until the desired overall look has been achieved.

This kind of "shadow design" is the professional approach to lighting design. The lighting designer is a shadow designer, treating the task as an artist treats the canvas, by painting gradations of darkness, using light sparingly on illuminated surfaces, thereby using a "whispering light" palette as the design tool.

Shadow design can be compared to oil painting techniques. French researchers who studied Leonardo da Vinci's paintings, including the *Mona Lisa*, to analyze the master's use of successive ultrathin layers of paint and glaze—a technique that gave his works their dreamy quality—found that da Vinci painted up to 30 layers of paint to meet his standards of subtlety. The technique, called *sfumato*, allowed Leonardo to give outlines and contours a hazy quality and create an illusion of depth and shadow.

Almost 100 years of the technical lighting design era have made high-glare and excessively illuminated city structures the standard for new lighting designs. The fascination of darkness is totally missing. The only way to succeed with such city structures is to dismantle old lighting installations and start from darkness. Some cities are considering such lighting renovations/innovations in certain parts of the city structure.

Studies on how best to meet the darkness play an important role when we select minimum lighting design values for lighting compositions in architectural outdoor lighting solutions. A certain face-recognition distance (3 meters) is mandatory for relaxed movement in an outdoor space, and that calls for certain lighting design values. With this "tool" we can avoid environments that are too dark.

It might be noted that shadow design has mystical connections with music; both require thoughtful composition to be enjoyable and successful.

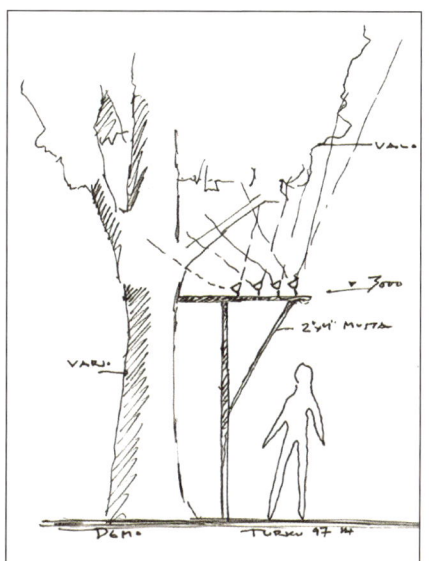 

Figure 58 left and right: Testing our new idea, using 400 cheap halogen lighting units and a wooden structure. The purpose of the demonstration was to study the total amount of light that each tree would need to be architecturally beautiful, low-key, sufficiently efficient and completely glare-free from all viewing directions and heights (bridges). The overall view was important. Viewed from one side, the tree trunk was light, and from the other side dark. This reflects life's dichotomies: Light-shadow / sweet-salty / love-hate / birth-death, etc. **Photo:** Aura River Report, final new luminaire design and solution designed by Honkonen & Oksanen

Figure 59 left and right: Customized solution: Simple black aluminum pole, steel lamp chamber with two 150W metal halide lamps and 100-degree optical output angle, strong glass surface on the lamp chamber. **Photo**: Aura River Report

Figures 591 left middle right: Nature takes care of changing the lighting to different times and moods. **Photos**: City of Turku

Tree lighting, molecular plant biology and light pollution

Tree lighting in general

In her 2017 blog post "Hit the Lights! Pollution's Negative Impact on Urban Trees,"[12] Nicole Peterson writes that artificial light "confuses trees by extending the day length, which can change flowering patterns and promote continued growth, preventing trees from developing dormancy that allows them to survive the rigors of winter weather."

She continues: "When artificial lighting is necessary, mercury vapor, metal halide, or fluorescent lamps give off less harmful wavelengths of light. Fixtures should be shielded so the light is directed on the ground, and not up into the sky, and uplighting trees should be avoided. Where possible, lights should be turned off or dimmed during off-peak hours, to give the trees a period of darkness. There is also some variation in the susceptibility of trees to lighting, as seen in Table 1. Using tree species that are less sensitive to artificial light can help with overall tree health and longevity. If you are planting trees where night lighting already exists, select those with low sensitivity to light to ensure the best possible outcome."

Table 1. Sensitivity of woody plants to artificial light

High	Intermediate	Low
Acer ginnala (Amur maple)	*Acer nigrum* (Black maple)	*Fagus sylvatica* (European beech)
Acer negundo (Boxelder)	*Acer palmatum* (Japanese maple)	*Fraxinus americana* (White ash)
Acer platanoides (Norway maple)	*Acer rubrum* (Red maple)	*Fraxinus nigra* (Black ash)
Betula alleghaniensis (Yellow birch)	*Acer saccharum* (Sugar maple)	*Fraxinus pennsylvanica* (Green ash)
Betula lenta (Sweet birch)	*Cercis canadensis* (Redbud)	*Fraxinus quadrangulata* (Blue ash)
Betula nigra (River birch)	*Cornus sanguinea* (Bloodtwig dogwood)	*Ginkgo biloba* (Ginkgo)
Betula papyrifera (Paper birch)	*Gleditsia triacanthos* (Honeylocust)	*Ilex opaca* (American holly)
Betula pendula (European white birch)	*Ostrya virginiana* (Ironwood)	*Liquidambar styraciflua* (Sweetgum)
Betula populifolia (Gray birch)	*Phellodendron amurense* (Corktree)	*Magnolia grandiflora* (Southern magnolia)
Carpinus caroliniana (Hornbeam)	*Quercus alba* (White oak)	*Malus sargenti* (Sargent's crabapple)
Catalpa bignonioides (Southern catalpa)	*Quercus rubra* (Red oak)	*Picea engelmanni* (Engelmann spruce)
Catalpa speciosa (Northern catalpa)	*Quercus montana* (Rock chestnut oak)	*Picea glauca* (White spruce)
Cornus florida (Flowering dogwood)	*Quercus stellata* (Post oak)	*Picea glauca densata* (Black Hills spruce)
Cornus sericea (Redosier dogwood)	*Sophora japonica* (Japanese pagoda tree)	*Picea mariana* (Black spruce)
Fagus grandifolia (American beech)	*Tilia cordata* (Littleleaf linden)	*Picea pungens* (Colorado blue spruce)
Liriodendron tulipifera (Tuliptree)		*Pinus banksiana* (Jack pine)
Platanus hybrida (London planetree)		*Pinus flexilis* (Limber pine)
Platanus occidentalis (Sycamore)		*Pinus nigra* (Austrian pine)
Populus deltoids (Cottonwood)		*Pinus ponderosa* (Ponderosa pine)
Populus tremuloides (Quaking aspen)		*Pinus resinosa* (Red pine)
Robinia pseudoacacia (Black locust)		*Pinus rigida* (Pitch pine)
Tsuga canadensis (Hemlock)		*Pinus strobus* (White pine)
Ulmus americana (American elm)		*Pyrus calleryana* (Bradford pear)
Ulmus pumila (Siberian elm)		*Quercus palustris* (Pin oak)
Zelkova serrata (Zelkova)		*Quercus phellos* (Willow oak)

Compiled from Cathey and Campbell (1975) and Hightshoe (1988)

Figure 60: Table 1

---

12 https://www.deeproot.com/blog/blog-entries/hit-the-lights-light-pollutions-negative-impact-on-urban-trees/

Biological research on illuminated trees on the Aura River banks in 2021

The following summarizes discussions between the author and PhD and docent Esa Tyystjärvi (Department of Biochemistry/Molecular Plant Biology, University of Turku, Finland):

Night lighting is known to affect many plants. The so-called short-day species require a long night to bloom. These plants may be prevented from flowering if electric lighting, such as street lighting, cuts the night too short. Long-day plants, on the other hand, bloom when the day is long enough. In agriculture, the day length requirements for special crops are carefully considered, and very often plants are intentionally grown at the "wrong" day length so that they do not flower. For example, letting a lettuce flower is not desirable if you intend to get lots of edible leaves.

A great many plants are day-neutral, meaning that they bloom regardless of the length of the day. The length of the light period often mainly affects the timing of flowering. For example, our most common test plant (in the university's research), the Thale cress, is a long-day plant, but if it is grown long enough for a short day, most individuals will eventually bloom.

Regarding the day-length requirements of the trees on the River Aura banks, we can only say that linden, elm, oak, ash, silver birch and *Betula pubescens* bloom in Finland in summer, so at least they cannot be very strict short-day plants. In Turku, these species also bloom every summer. Shortening the night of a day-neutral or long-day plant is unlikely to have much of an effect on its flowering. However, to the best of our knowledge, no studies have been conducted on the trees on the Aura, so we cannot be sure. The Turku city gardeners know best about the day-length requirements for park tree species and varieties planted along the river.

The length of the day also affects how resistant some perennials are to cold in the autumn. However, a 30-year field experiment on the Aura's banks seems to show that the species growing there now cannot be very light-sensitive in this respect. Autumnal foliage senescence is a phenomenon regulated by both temperature and day length. It is also entirely possible that electric lighting will affect the timing and even the colours of autumn leaves.

Light acts in plants not only through photosynthesis, but also through several cytoplasmic photoreceptor mechanisms. Phytochromes are responsible for the red and invisible so-called long red responses, cryptochromes again for blue light responses. Flowering and many developmental phenomena are mainly regulated by phytochromes, while blue light affects, for example, the morning opening of air gaps and the movement of green particles within leaf cells. The light intensity required by cytoplasmic photoreceptors is typically quite low, which is why so-called white light usually contains enough blue light, for example. These reactions are not on/off things. Thale cress grows quite well, for example, at wavelengths of 440, 470, or 660 nm (in rather narrow wavelength bands), even though it lacks red or blue light.

When assessing the effect of light, its intensity (in terms of photochemical reactions, preferably the density of the light flux) is relevant. Moonlight, of course, has been studied in this regard because the full moon can in principle be bright enough to disturb the nocturnal dormancy of short-day plants, and the closure of some plant leaves to the "sleeping position" has been proposed as an adaptation to avoid moonlight. This phenomenon, known as "nyctinasty," only very rarely occurs in public park trees, but it is not impossible.

The first sentences of Dr Tyystjärvi's research report provided clear guidance for the tree lighting design team: "It is essential to establish whether the illuminated trees are long-day species or short-day species."

According to horticulturalist Aki Männistö:

"The illuminated trees of the Aurajoki project date from the 1860s. Their condition is good/moderate. So not excellent, but not bad either. The lighting has not affected these trees in any way. The only risk in wood lighting cases is in digging cables. In other words, when the lighting units come under the tree foliage and close to the trunk, digging to lay their cables is a critical issue."

Research in the River Aura Tree Lighting case showed that trees that had been illuminated from below for over 25 years, night after night, had not suffered at all. The bigger problem in this case was that the style of lighting was vainly illuminating space. (New nocturnal lighting methods that accommodate both architecture and biodiversity are introduced in the section titled Biodiversity-friendly Tree Lighting Solutions.)

Light pollution and the Aura River case study in 2021

Although the River Aura trees have fortunately shown themselves resilient, the negative effects of light pollution on many species have been revealed in studies. The lighting of Turku's River Aura, where trees (oaks) have been continuously lit at night since 1998, is an important object of research. The total lumen flux of the project is over 2 million lumens aimed straight up at the trees/sky. A heuristic calculation shows that an equal amount of light pollution is produced by 100 km of street lighting surface, depending on the street structure, the street lighting classification used, and the level of operating technology (tree lights constantly shoot 100% lumen output every night).

Heuristic mathematical calculus of light effect to the sky (and birds):

Figure 61: In clouds containing a lot of water droplets/particles, each beam of light (in the case of the River Aura, from a hundred tree illuminations) is scattered and continues its journey as "new" rays of light. The reaction recurs almost indefinitely. Photo engineer Olli Oksanen

   This overall scattering of River Aura light results in a large and dense veil of light in the sky. **Photo:** bird's eye view by drone, by engineer Olli Oksanen. (Clear evening, no clouds, no water droplets in the air, no scattering.) Disturbing light beams aimed toward birds' eyes persist from all viewing angles. It's important to recall that birds' vision is on average 10 times more acute than that of humans, which makes the glare from this much more disturbing to birds than to us. A bird's eye is also more sensitive to the broad spectrum of light.

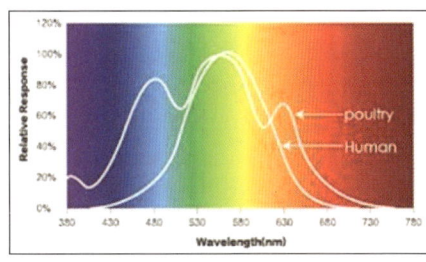

Figures 62 left and right: Left: "Eagle Eye" by Peter Kaminski is licensed under CC BY 2.0. Right: Eye sensitivity curve comparison between human and poultry. Photo: CLSP

Figures 63 left and right: Honkonen & Oksanen. This unique tree lighting unit is an architecturally anonymous, inconspicuous element within the historical setting of the River Aura. The luminaire is technically glare-free from all human viewing angles. The exception is the bird's eye view, where light is emitted in a wide beam, disrupting birds' lives and producing more than two million lumens of luminous flux in the area indicated by the calculation in cloudy weather, at an altitude of 200 m. This can disturb migration at an altitude where migratory birds fly in cloudy weather.

In 1997, nobody gave any thought to bird migration routes for this project. But we have to start taking biodiversity elements into consideration. Add to that problem the issue of obscuring our view of the stars with this veil of illumination and continuing unnecessary energy consumption. A better solution can be found when designers are vigilant and aware of the effects of their work on biodiversity. Education is needed, and the design team absolutely should have included a biologist. Photo: Left Vesa Honkonen, right: Photo: Bird's eye view by drone, Olli Oksanen.

Calculations:
- Total lumens from 100 tree illuminations: **2,102,500 lumens.**
- Uniform light-veil extent and average illuminance value in a cloudy light-veil at an altitude of 200 m (migrant bird flight altitude in cloudy weather): **Eav. = 5,82 lx.**
- Total lumen flux penetrating into the clouds: **1,425,900 lumens**.
- A horizontal surface area (500m x 700m) facing light at the bottom of a hypothetical cloud at 200m above the ground.

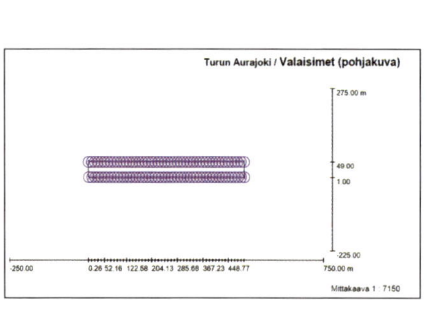
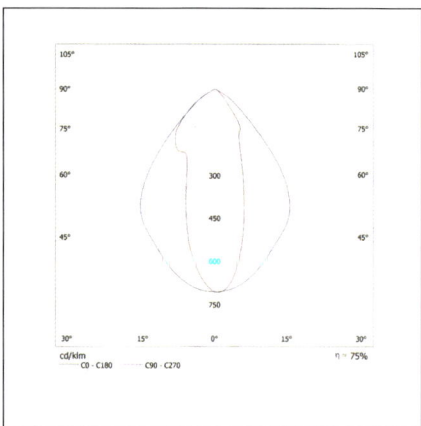

Figure 64: Left; calculated area, right; simulated luminaire light output. **Photo:** Engineer Pauliina Oksanen

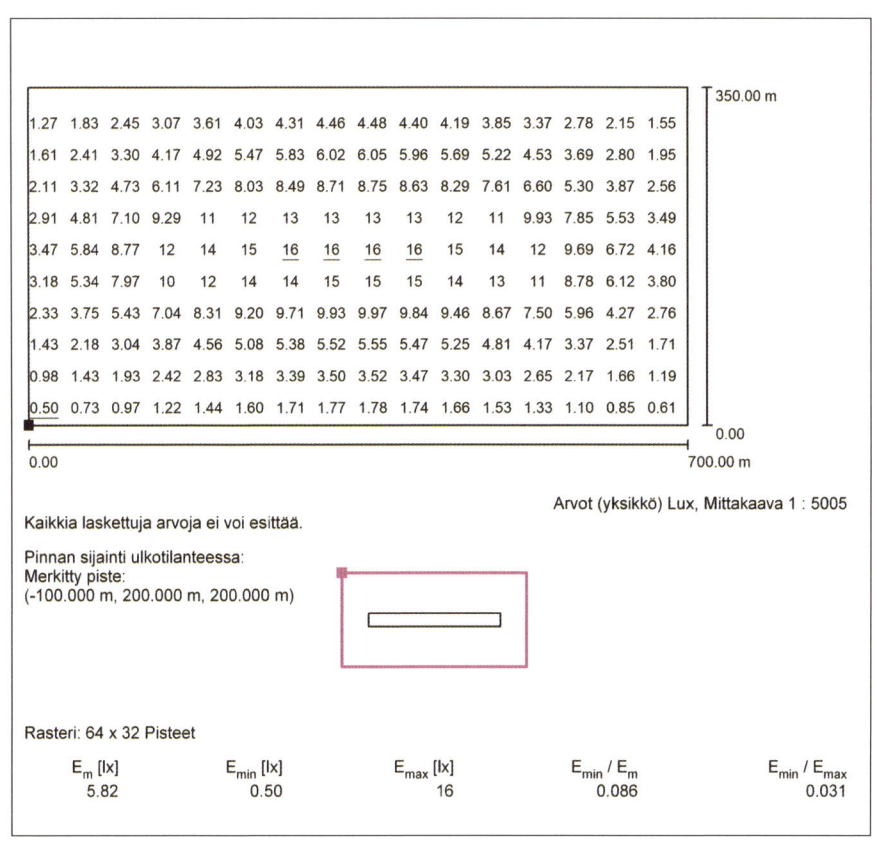

Figure 65: Numerical Illumination values on the cloud surface 200 meters high.
Photo: Engineer Pauliina Oksanen

The light veil's impact on biodiversity: lost night and disturbed migration

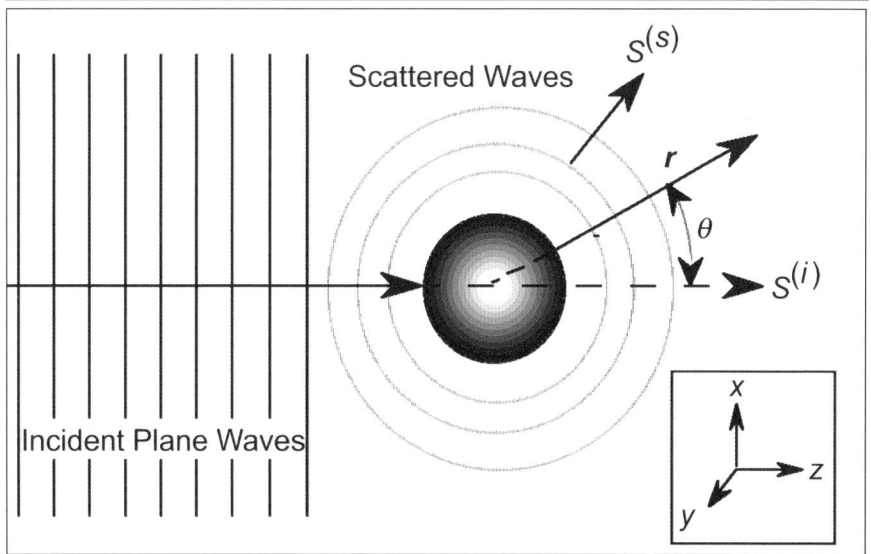

Figure x: Light scattering

A veil of light is produced in a cloudy sky by the process of "scattering." *Scattering* refers to the reflection of a beam of light in aberrant directions. In scattering, an atom absorbs a photon, or particle of light, and emits it immediately, but usually in directions different from the original. The radiation appears to be reflected from the medium. Scattering also dims the beam of light passing through the atmosphere.

Scattering is a physical process in which radiation, such as light, changes direction when it encounters an obstacle or a change in the density of the material in which it travels. We can understand the effects of this by using Mie theory, an analytical solution of Maxwell's equations which calculates the interaction of electromagnetic radiation with a spherical body. Mie scattering theory, is used, for example, to calculate the scattering of visible light from atmospheric water droplets. In clouds with a lot of water droplets, the scattering rays caused by each light beam (in the River Aura case, from millions of light rays shooting into the sky from a hundred lighted trees), continue their journey as "new" scattering light beams. The reaction repeats almost indefinitely. This overall scattering results in a large and dense veil of light over the river.

Figure 66: Example of light pollution scattering from water droplets in clouds. Left: The starry sky during a power outage. Right: Light pollution obscures the stars before the power outage. Photo: "Light pollution: It's not pretty," by jpstanley, is licensed under CC BY 2.0.

How the light veil disturbs migratory birds

Figure 67: Migratory birds navigate with the help of a magnetic field formed by the poles and the "compass" of the right eye. In cloudy weather, migratory birds fly below the clouds. This allows the light veil and the light distribution surfaces of the lights to entice the birds away from their straight path, to orbit the light, with possibly serious and devastating consequences. Photo: Finnish Nature Conservation Association

 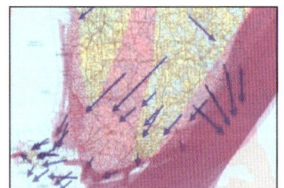 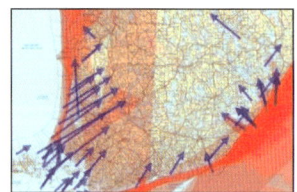

Figures 68 left middle right: Composite map of the main migration routes of about 20 bird species over the Turku city area. Many of these migratory bird species are also nocturnal migrants, among them arctic waterfowl, loons, cormorants and sea eagles. These species are threatened by light pollution in both spring and autumn. Left: City of Turku. Middle: Autumn migration. Right: Spring migration

Biomimicry and biodiversity-friendly tree lighting solutions

Biomimetics refers to the technological imitation of mechanisms occurring in nature and the social behavior of various organisms. By emulating nature's models, systems and elements, biomimetics (or biomimicry) can often help solve complex human problems.

A famous example is Leonardo da Vinci's flying ships, in which the flight of birds was modeled for engineering sciences. About 600 years later, in 2004, biomimetics at Penn State University designed shape-shifting airplane wings that morphed according to the speed and duration of flight. The models were obtained from various bird species whose flight speed depended on their shape. In the invention, the wings are covered with fish-inspired "scales" that slide over each other in a chosen way. Another famous example of biomimetics is Velcro, invented by Swiss engineer Georges de Mistral, who was inspired when he was cleaning his dog's fur of burdock seeds whose "hooks" stuck stubbornly to the fur.

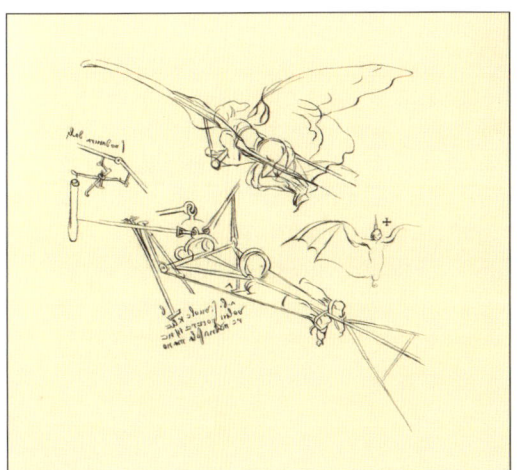

Figures 69 left and right: Left: "Leonardo da Vinci: Diagram of a proposed flying machine (1789)," from Toronto Public Library Special Collections, is licensed under CC BY-SA 2.0. Right: "Leonardo Da Vinci's Helicopter invention design," by Ell Brown, is licensed under CC BY 2.0.

Bioluminescence simulation options

One professional biomimetic solution to rid ourselves of the light pollution veil in the sky involves simulating bioluminescence, a natural production of light in certain species.

Perhaps the best-known form of bioluminescence is that of some types of phytoplankton, drifting marine organisms. Tiny, unicellular marine plankton known as dinoflagellates use bioluminescence as a form of self-defense, producing a light flash that lasts a fraction of a second. This type requires water and won't be useful in tree lighting projects.

However, there may be a brilliant alternative: bioluminescent fireflies.

Bioluminescence of fireflies

This type of bioluminescence has real possibilities for being emulated in tree lighting projects. According to the article "Chemistry of Firefly Bioluminescence,"[13] by Bruce R. Brancini of the Department of Chemistry at Connecticut College: "Bioluminescence is an enchanting process in which living organisms convert chemical energy into light. Light is produced by the oxidation of an organic substrate, luciferin, which is catalyzed by an enzyme called luciferase. Nature has an amazing diversity of light-emitting organisms, including bacteria, fungi, crustaceans, mollusks, fish and insects (Hastings, 1995). Although the specific biochemistry of bioluminescence is different in different species, all involve an enzyme-mediated reaction between molecular oxygen and an organic substrate. It is also likely that all bioluminescence processes involve the formation and degradation of a four-membered ring of peroxide or linear hydroperoxide (Wilson, 1995; Wood, 1995)."

---

13 http://photobiology.info/Branchini2.html

Figure 72: "Firefly-scooter-travel," by Sinchen.Lin, is licensed under CC BY 2.0.

Fireflies have several reasons to shine. When they are small, they shine briefly and mostly at night. Predators avoid fireflies because they produce self-defensive steroids in their bodies, and glowing is a sign that they are repulsive. As adults, different species of fireflies emit characteristic and unique flashes of light to identify their own species-specific sex. Female fireflies may choose their mates based on the male's flash patterns, preferring more frequent and higher intensity flashes. Not all species of fireflies produce light; some use pheromones as chemical messengers to potential mates, while some use both pheromones and light.

Figures 73 left and right: Each species of firefly produces its own characteristic color of light. Some glow yellow or orange, and others glow blue or green. Tiny, efficient LED bulbs work well for firefly simulation purposes (e.g. blue/green and red/orange), with a luminous Intensity around 4000~20000mcd. **Left:** "Glow in the dark firefly lantern," by Backyard Boss, is licensed under CC BY 2.0. Right: "Firefly," by Adam Greig, is licensed under CC BY-SA 2.0.

A firefly bioluminescence simulation
Rendering of the desired atmosphere

Figures 74 left and right: Left: "Forest Fireflies," by Brett Jordan, is licensed under CC BY 2.0. Right: "Fireflies and Star Trails No. 3," by ikewinski, is licensed under CC BY 2.0.

When dusk spreads its veil, this forest, harmoniously and elegantly lighted by nature, forms a glare-free and intriguing atmosphere, no matter what direction you approach it from. Entering it, one has the feeling of stepping into a different world, a fantasy world, poetic and mystical. One of the best lighting concepts in urban parks is to achieve such biomimicry.

Artificial flickering fireflies can evoke memories of days gone by in our brains, recalling, for example, the romantic notion that poor students once gathered fireflies in jars so as to read by their light. Our experience and many demonstrations show that tree lighting needs no more than this to be effective. If illuminated this way instead, the large, festive trees in Turku's Aura River park would take the leading role in creating a romantic ambience. In this style of lighting, we see no animals except fireflies, but we can potentially hear them; adding other animal sounds would be an interesting addition, reinforcing the biomimetic theme. Designers could consider using frog, grasshopper and waterfall sounds in a similar way.

Figure 75: An experiment

This concept has been successfully tested by the author's team and received with delight in several environments, although it has yet to be accepted as a replacement for the Turku lighting scheme. Therefore some imagination is required to see it as applied to the River Aura setting:

Twilight is falling on the Aura River; everything is calm and quiet, but energetic plants are still growing and expanding during the night. You can almost sense the growth of nocturnal Nature. It differs from the hectic daylight growing time. We are touched by the power of life that dominates this place, not only of growing plants, but also the power of light. The atmosphere produced by the fireflies' twinkling lights is so beautiful, warm and inviting, its greenish glow reminding us of the green life around us, the dance of male and female ... the power of life itself.

Technology for the artificial firefly simulation

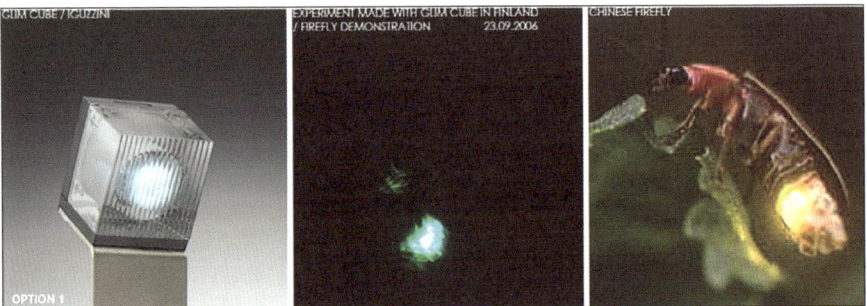

Figure 76

Artificial fireflies: the firefly stimulation used 2000 Iguzzini´s dimmable 1 W Glim Cube luminaires. 1000 in amber and 1000 in green. To convincingly stimulate fireflies, we had to be able to operate them flexibly, dimming, shutting down completely and flickering. This all is possible with Iguzzini´s operating system. Glim Cube needed to be modified, so that every side was illuminated so there was a need for customizing to get rid of the riffling.

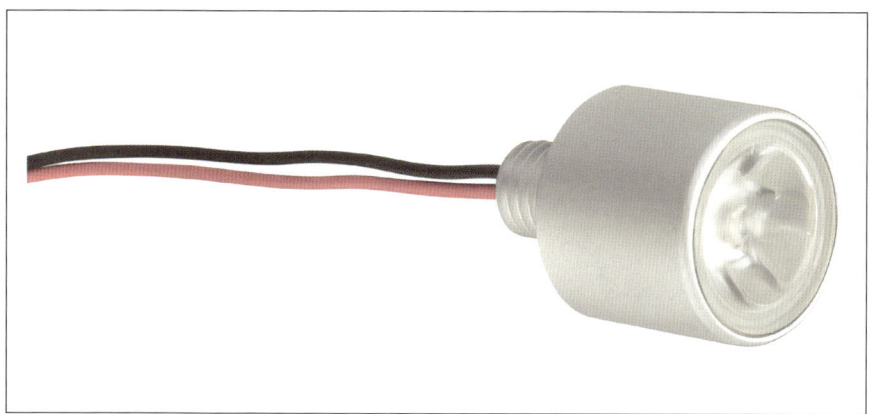

Figure 77: This LED unit requires a custom-made light diffuser in front of the light distribution surface.

LED technology has evolved tremendously since 1997. LED lights in thousands of colours and sizes are now available, enabling a perfect natural firefly simulation. I invite all potential "firefly designers" to demonstrate for themselves how an LED simulation can imitate a real firefly. A large lighting project like that along the River Aura involves thousands of LED units, so it's important that the simulation LED unit is perfect for the purpose.

Tree lighting that simulates fireflies must be planned carefully if the planned life cycle of the project is long. Success requires the careful cooperation of a biologist, an architect, an electrical engineer and a darkness designer. The design package needs to include the exact design of the system and all the details. Versatile demonstrations help to guarantee the success of the project.

An example of a power supply system structure:

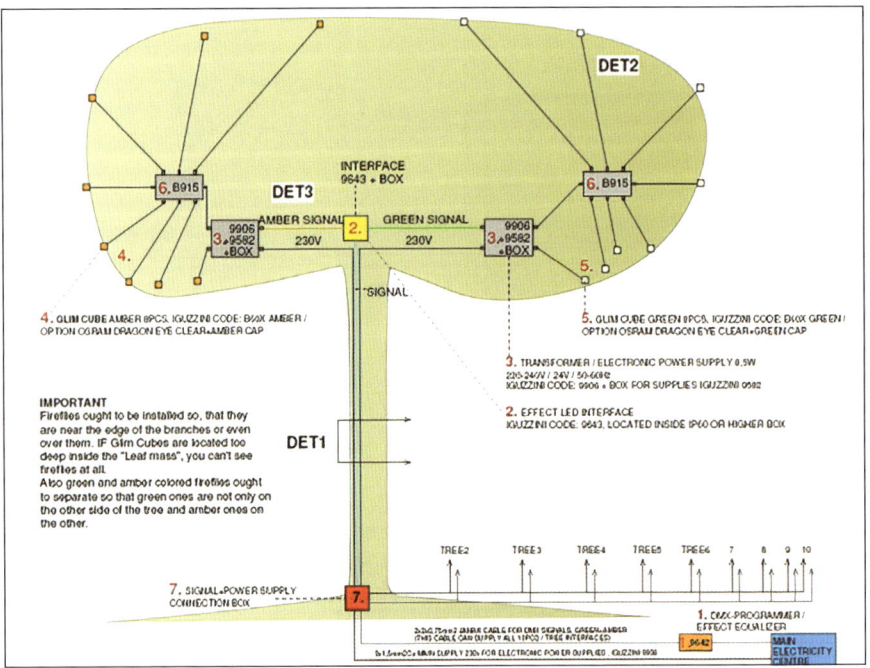

Figure 78

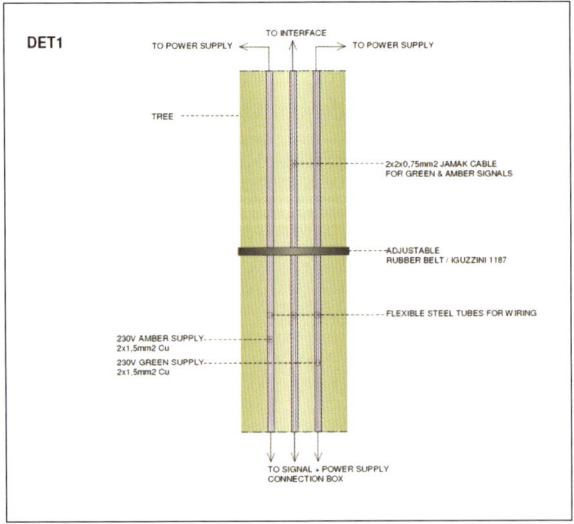

Figure 79

90 | DARKNESS DESIGN AND BIODIVERSITY

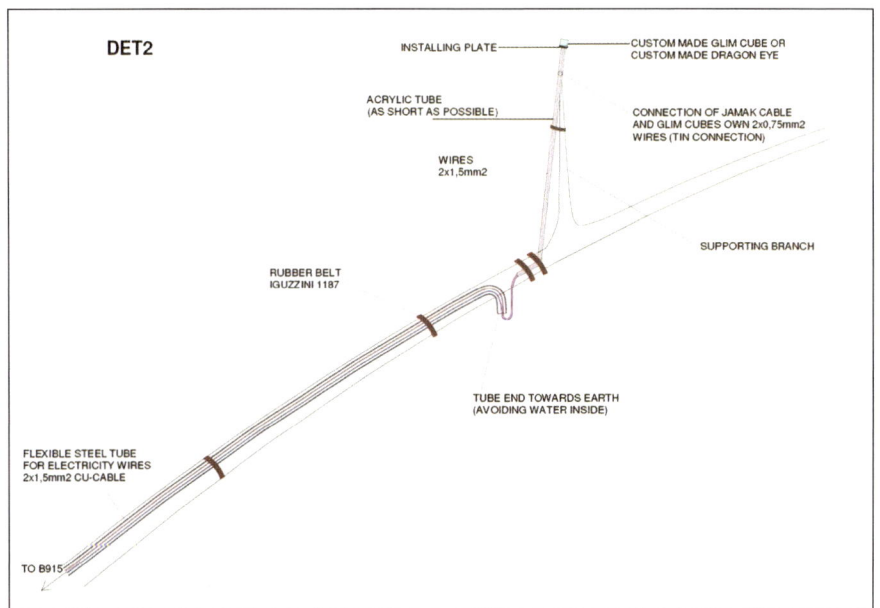

Figure 80

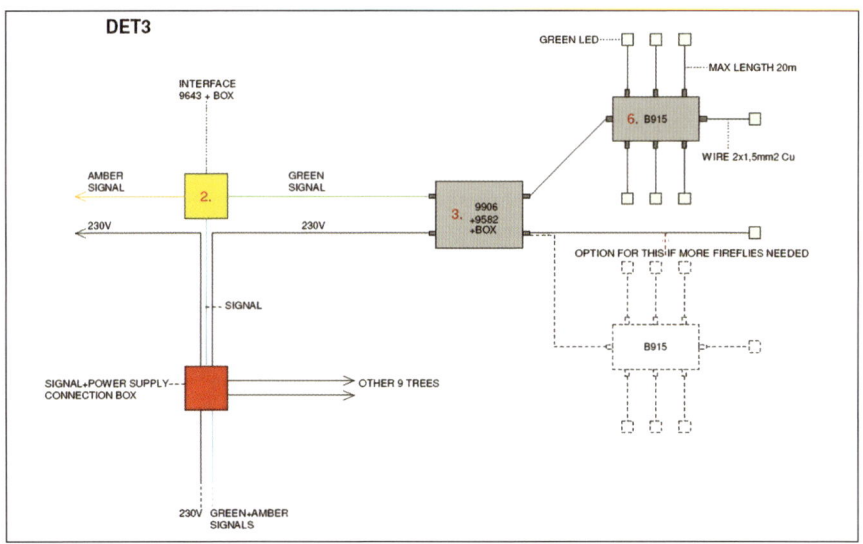

Figure 81

CHAPTER 2: THE TECHNICAL OUTDOOR LIGHTING PARADIGM | 91

A temporary option

Figure 82: "Manmade fireflies," by Fekaylius, is licensed under CC BY 2.0.

Alternatives to bioluminescence

If the bioluminescence solution proves to be too difficult to design or implement, lighting the tree from above is an alternative. Though less than ideal. Illuminating from above is a less light-polluting way than from below. It imitates the natural filtration of sunlight from the tree's upper branches to the ground. A practical demonstration is always particularly important in such cases, depending on forest size, sizes of trees to be illuminated, and other environmental elements. The demonstration will check the appropriate intensity of the light, how to conceal the light-distributing surfaces from all possible viewing directions, and the quality of the assembly.

Figure 83: Photo "Angel of Light" from Collection of Amazing Sunrays in The Forest In The Netherlands. Photo: Photographer Edwin Mooijaart

Aura River tree lighting: comparing possibilities in 1997 and 2021

Complete firefly simulation is now easier to implement than even a decade ago. It is important that the color used in the simulation matches the color of a real firefly, and that the "flicker simulation" effect matches the flicker of a real firefly. "Normal" electronic lighting control systems are not fast enough to simulate firefly flicker. But the DMX control system used in theatrical lighting is perfect for this purpose. A practical example can be viewed online. A projector has been used as the lighting technology in the online DMX demonstration, but this gives an indication of how smooth effects are achieved with this control system.

Benefits of imitating bioluminescence in the Aura River project

The energy consumption of one bioluminescence unit (eg Iguzzini Glim Cube): is 1.2W, and lumen output 80 lumen. The Aura River trees have 20 units per oak tree, which equals 24W per tree. So 100 oaks consume 2.4 kW. (Compared to the 30 kW consumed by the 1997 installation.) The total light output of the bioluminescence solution is (without light direction clarification) 2000pcs x 80lm/pcs = 160 klm. (Old installation 2102.5 klm.)

In the current installation, light is projected directly to the sky. In the proposed firefly solution, the LED unit output stays within the angle of a designated space, not going straight to the sky. The estimated light pollution produced in the sky is less than 3% compared to the old lighting. Light pollution by bioluminescence imitation is negligible. The solution is also fascinating, especially if you add natural sounds (frogs, flowing water, birds, etc.). The solution integrates nicely with nature and imitates biodiversity. Some comparisons:

Challenges and tasks for the lighting collaboration team:

Architect: Visual responsibility.
Lighting/shadow designer: Artificial firefly simulation, firefly LEDs, producing the final lighting design report.
Electrical engineer: Lighting control system design. (DMX control system, used programs and wiring systems, etc.).
Biologist (e.g. person from Birdlife International): Consider trees and light pollution, bird migration periods, lost night, firefly flickering periods, etc.
Horticulturalist (e.g. Hortonomi Männistö in hypothetical River Aura project): Consideration of tree health.
Owner: Calculation of costs and payback time.
City planning official: Present the City context.
Conclusions drawn from this alternative to upward-aimed lighting
- Light pollution by bioluminescence imitation is negligible
- The solution is fascinating, especially if you add sounds of nature (frogs, flowing water, birds, etc.)
- The solution integrates nicely with nature and imitates biodiversity
- Never illuminate trees upwards

UPWARD-AIMED GENERAL LIGHTING: LARGE, EFFICIENT LIGHT POLLUTION PRODUCERS

Illuminating space in vain

The close, 100-year collaboration between lighting manufacturers, electrical engineers and architects has produced tens of thousands of products for the market. Unfortunately, few have considered the impact of light pollution on biodiversity. Only in recent years has the issue has begun to be considered at some modest level. Changing a century's worth of implementation is difficult, but there is no need to lay blame. Cooperation between different fields has always been difficult. If it had been easy and natural, we would not have climate change and other biodiversity issues.

In this section of the book, we'll discuss proposed changes to large glass elements to avoid light pollution and direct light output to the sky, using as an example the design and implementation of lighting for the light prisms at Helsinki-Vantaa Airport. This project was awarded "Lighting project of the year" by the Illuminating Engineering Society of Finland in 1997. It sets an example of innovative and biodiversity-respecting change at minimal cost and offers a model that can be applied to many existing projects with large light elements and large light ceilings, as well as guidance for new projects.

Helsinki Airport Light Prisms (1997)

Helsinki-Vantaa Airport has the largest luminaires in the world. Seen from the outside, the glass lanterns appear to be large blocks of ice that glow in the Nordic night, and inside the Gateway area they create a harmonious impression as massive light-reflecting elements. Each glass lantern weighs 2200 kg. The light-producing fixtures for the glass lanterns are located on the roof. Their light is directed through glass side faces to adjustable aluminum reflector wings, which in turn reflect light into the interior of the airport.

The light production from glass lanterns has been implemented using the applied lumen method. Mathematical formulae and calculations are presented to illustrate the more technical aspect of lighting design.

Architectural design: Pekka Salminen Architects Ltd. Lighting design: author

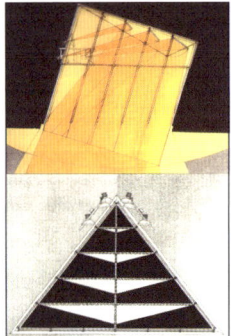 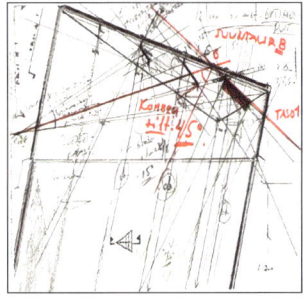 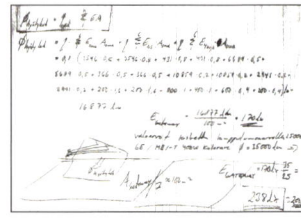

Figures 84 left middle right : This image gives a hint of the style of calculation. The actual lighting plan included 10 pages of manual and computer calculations. Photos: Pekka Salminen Architects Ltd and author

Figures 85 left and right: Project is ready. Left: Snow Lanterns, Ice Boulders, Ice Cubes on the roof of the airport. Right: Eternal lighting/daylight + electric lighting. Photos: Pekka Salminen Architects Ltd.

Though this lighting solution is successful artistically, it causes light pollution, especially in foggy weather. In the vicinity of airports, light solutions in a symmetrical line should generally be avoided if they can be combined visually with runway guiding lights.

After the project was completed, I talked to one of the flight captains and asked if the light prisms bother him during the landing phase. The answer was surprising. He thought the solution was good because he is landing his plane with manual control with the help of these light prisms, and never wants to use the autopilot.

Light-pollution free "Ice Boulders" (a renovation project and model for new projects)

 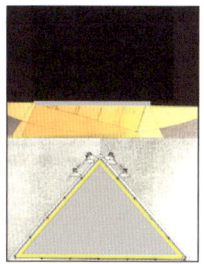  

Figures 86: Left, second left, third left, right
Figures: A simple and workable solution for the light-pollution issue in this case is to use light prisms only as daylight producers during the day. During dark periods, automatically operated ceiling-colored roll-up blackout curtains (gray in the picture second left) could act as reflectors of interior lighting while preventing electric lighting from escaping into the sky. LED strips (yellow stripes at second left) could be installed to the steel support structures. Blackout curtains would thus act as large, triangular light-distribution surfaces and the glass elements would be still beautifully visible (photo right). The 120 existing floodlights (photo left) could be removed as superfluous, resulting in a biodiversity-friendly, light-pollution free lighting solution (photo 3rd from left). The renovation can earn LEED points with the choice of an environmentally friendly curtain material in addition to the new anti–light pollution philosophy. Photos: Modified by author from photos by Pekka Salminen Architects Ltd.

Figure 87: "File: Das Reichstagsgebäude – The Reichstag building - Berlin – panoramio (cropped).jpg" by Jens Cederskjold is licensed under CC BY 3.0.

The Reichstag building, Berlin

The glass dome of the Reichstag building in Berlin, Germany, designed by architect Norman Foster. The low-key evening light of the huge glass dome in the picture sits on the cityscape and promotes biodiversity with its light pollution-free status.

Figures 88 left and right: Left: "Reichstag dome building in Berlin" is marked with CC0 1.0. Right: "Reichstag building in Berlin, Germany" is marked with CC0 1.0.

Challenges and tasks for the lighting collaboration team

Task: Produce a biodiversity-friendly and light-pollution free lighting solution for large glass elements.
Architect: Visual responsibility.
Lighting/shadow designer: Design a light-pollution free lighting solution, producing the final lighting design report.
Electricity engineer: Lighting control system design
Biologist (e.g., person from Birdlife International): Consider issues involving light pollution, bird migration routes and periods, lost night, etc.
Glass consultant: Glass materials and light behavior.
Owner: Calculation of costs and payback time.
City planning office official: Present the City context.

ROAD/STREET/VEHICLE LIGHTING

Lack of cooperation and a great opportunity

A problem that contributes to the huge amount of light pollution contributed by street lighting is the lack of collaboration between experts in road lighting and those dealing with vehicle lighting systems. We'd like to inspire international stakeholders in both fields to work together to combine their expertise. Global pooling of research in this area could have major biodiversity-saving effects. Vehicle lighting has developed with the same technical elements as street lighting (multifunctional LED lighting with electronic elements). Involving expert biologists in the study would boost motivation on both sides.

Over a hundred years of research in road and street lighting has focused on developing technical arguments that have ignored the lighting properties of vehicles. LED technology and the use of electronics in street lighting control systems and even possibilities for integrating street lighting with other systems in urban structures have developed in unprecedented ways. That's why it's absolutely essential to take modern vehicle lights into account in future road lighting. Pondering whether street lighting might actually become less necessary,[14] world-renowned lighting expert John D. Bullough relates having the opportunity to to

---

[14] Bullough JD. "Opinion: Will Road lighting wither?" Lighting Research & Technology. 2017;49(6):672-672. Doi:10.1177/1477153517730723"

drive a vehicle equipped with ADB headlights under evaluation by the Lighting Research Center.

"I expected sluggish, distracting changes in the beam pattern, and I anticipated road rage from all the other drivers I'd be glaring. I was pleasantly surprised by the system's robust performance. Further, our evaluation revealed that the system can offer tangible safety benefits: high-beam forward visibility with low-beam glare. Indeed, ADB systems work so reliably that there may be little reason to limit them to conventional high-beam intensities, since they can rapidly reduce intensity (and glare) that would be observed by others."

An important collateral benefit of ADB headlights is that drivers could use high beams more frequently. Even on rural roads, nighttime drivers use high beams only 25% of the time, unsafely using their low beam headlights for the remaining 75%. This could have implications for road lighting, since standards state that reasons to install lighting include reducing headlight glare and supporting vision beyond the reach of headlights. To the extent they sustain driver visibility without increasing glare, ADB headlights might permit road lighting to wither, at least in certain situations and locations.

"Undoubtedly, much research and demonstration work will be needed to confidently reach this conclusion. But less dependence on road lighting could have other benefits, like less conflict between road and vehicle lighting, reduced energy use, fewer poles, and improved connections to our nighttime sky. Are they worth the effort?"

Calculations presented later show the biodiversity-balancing effects that can be achieved by combining electronic control systems for road and street lighting and LED technology with intelligent vehicle lighting technology. Prior to these technologies, the discharge lamp lighting solutions were not adjustable at a reasonable cost for road lighting. Experiments done as early as the 1970s and 1980s proved futile.

Glare and high energy loss are only one consequence of the light pollution caused by road lighting. Larger impacts include disturbance of wildlife and destruction of biodiversity. Further down, we'll look at the results of a biological study of light pollution.

Obstacles to cooperation, and some solutions

Human barriers to cooperation on road lighting

Why is it that road and street lighting experts, architects working on urban structures, and biologists who have studied biodiversity and light pollution have failed to work on these problems cooperatively? As mentioned earlier in this book, Rosabeth Moss Kanter, a professor of economics at Harvard University, suggests some reasons people typically oppose change.[15] I'll reframe those below in the context of street lighting and light pollution. In each case, I've suggested ways to overcome these psychological barriers.

Reason 1: Loss of control

In the area of road lighting, the challenge is to apply biodiversity-friendly design methods, moving from lighting design to skillful shadowy nocturnal design, without compromising safety. At the same time, care must be taken to minimize reflection of light from road surfaces to the sky and deep into forests. This requires collaboration between researchers in road lighting, vehicle lighting, and biodiversity. This kind of collaboration will seem odd and disruptive to the professional autonomy of the experts in these three sectors.

But rather than focusing on who wields power, we need to take into account the need to preserve biodiversity and reduce energy needs, while maintaining safe and comfortable driving conditions. Our concern for professional self-determination can be a barrier to change. But Nature cannot speak for itself. It's vital to generate multidisciplinary cooperation so that soon we will all can enjoy glare- and light-pollution free, contrast-rich nocturnal spaces, even on our roads and streets.

Worrying changes in biodiversity and climate (rising temperatures, floods, drought, an alarming decline in insect populations, light pollution, etc.) have affected everyone in the world, and attitudes have moved towards demanding mandatory measures to save biodiversity and reduce energy consumption. Now is the right time to change our philosophy of lighting as well, by reimagining the methods and skills needed to meet nocturnal design challenges—working cooperatively instead of focusing only on our personal proposals to tackle the identified problems. We could all learn from each other's expertise and find the right path together.

---

15   Harvard Business Review (September 25, 2012

Reason 2: Excess uncertainty

People are often reluctant to venture into the unknown, even though it promises a better life. The transition from road lighting design to shadow-rich nocturnal road Darkness Design requires hard work and strong motivation. To overcome inertia, we need a sense of security and an inspiring vision. The point of this book is to create certainty about a process that's new to the profession.

Confidence will come with a better understanding of two things: the remarkable advances in lighting technology that make more biodiversity-friendly solutions in road and vehicle lighting possible, and the usefulness of cooperation between people working in different fields. Both are motivations to move into a new age of cooperative activity.

Reason 3: Surprise, surprise!

Faced with having new ideas forced upon us, we typically react with resistance. If we have had no time to get used to the idea or prepare for the consequences, it's easier to say no than yes. We need to motivate experts in the fields of road lighting, vehicle lighting and biodiversity research to explore new Darkness Design methods and manufacture products that will minimize light pollution in old projects and eliminate it in new projects.

Unfortunately, there's no time to dally. Taking biodiversity into account in Darkness Design recommendations based on purely technical arguments may lead to anxiety and a refusal of cooperation. But adapting lighting design recommendations to Darkness Design will require professional cooperation, a positive attitude, and seamless collaboration between architects, engineers, and biologists to eliminate light pollution and glare and achieve safe and well-designed nocturnal driving spaces.

Focusing on a clear, common goal—Darkness Design to save biodiversity—is the way to overcome these old barriers. The decision to start using one's own professional know-how to work with others as a team on this clear goal can be a motivator greater than the fear of working with scientists and professionals in completely different fields.

Reason 4: Everything seems different

Change is meant to bring something different, but how different? We humans typically live by routines. Changes in those routines raise our awareness and push us into consciousness, sometimes in uncomfortable ways. Too many differences can be distracting or confusing. Changes in design are thoroughly justified by research findings from professional biologists and institutions like the Leibniz Institute. But the transition from lighting design to nocturnal Darkness Design brings up new and confusing issues.

Everyone working on a demanding design task needs to understand the absolute importance of the change. Associations or other interest groups, for example, might seek out motivated professionals interested in collaborating with "future groups" that respect and want to apply new technologies to save biodiversity. A group of "road and vehicle lighting specialists + biologists" could be just one example.

Reason 5: Loss of face

Those deeply invested on how things have always been done are likely to defend the status quo. A major shift in strategy brings up the uncomfortable notion that those ways may have been wrong. This automatically leads towards hardened positions. But the standard paradigm in road lighting has been in place for too long. This could be true of all professional groups (lighting, biology, architecture, engineering, etc.). We are just human beings. But because change is the one constant in life, we have to try to change our working style in a less flawed, cooperative direction.

Understanding that the world has changed makes it easier to let go and move forward and enjoy new challenges, a new era, new motivations in our design activities, creative freedom. The tools are there already; we need only to combine the research results and know-how.

Reason 6: Concerns about competence

Can I do it? Change often meets with resistance because it makes people feel powerless in the face of new challenges. For example, road lighting designers might express skepticism as to whether new Darkness Design methods work. Do we really need to work with vehicle lighting experts and with biologists? Deep down, they may worry that their skills will become obsolete.

A comprehensive rethinking of road lighting projects requires a wealth of motivation, cooperation, information, training, mentors, and support systems. New designs can be based partly on existing lighting design arguments. Concepts, master plans, and detailed designs are likely to change, but computer programs, technical mathematical formulas and the results of technical research can possibly be used as before. They can also be modified to suit the new Darkness Design methods, even if the outcome of cooperating means that road lighting is no longer necessary and that vehicle lighting proves able to ensure safety and comfortable driving while helping to conserve biodiversity in an unprecedented way. Road lighting and its design is not an absolute value. Many factors can facilitate change and promote collaboration among biologists, architects, and engineers.

Reason 7: It's more work

It's a universal challenge: changing from excessive illumination back to reasonable darkness with safety and enjoyable, glare-free lighting will indeed take a lot of work and disruption, which can be overwhelming for those closest to the issue. A decades-long replacement of massively increased lighting with professional, environmentally friendly Darkness Design methods will require a huge amount of work from the entire lighting field. Reforming lighting technology and arguing for simultaneous decisions by influential international institutions and stakeholders to save the climate and biodiversity will increase the work anyway. These elements ought to be viewed as inspiring, not as a burden.

Summary: Overcoming the human barriers to change

Change is the only constant in life, but humans are hard-wired to oppose change. Accomplishing a needed change requires leadership at all levels. A good example is the European Climate Law proposal, which aims to write into law the goals set out in the European Green Deal, to ensure that Europe's economy and society become carbon-neutral by 2050. (From the EU official website, **ec.europa.eu**).

A legacy of roadway lighting

Figure 89: The American Electric Light Tower (San Jose, 1885). With the advent of electrical lighting, the possibility of obtaining illumination levels similar to those of daylight prompted the question of how much technical effort it was feasible to invest. At the end of the 19th century, one attempt at providing street lighting was to mount floodlights on light towers. Glare and harsh shadow caused more problems than it solved, and this form of outdoor lighting was soon abandoned in San Jose. Photo: Wikipedia (Terms of use: Free) / Author's dissertation ISBN 978-952-60-7391-0 (pdf)

Figure 90: Lamplighter lighting a gas streetlight in Sweden, 1953. By this time, gas lamps had become rare curiosities. Photo: Author's dissertation ISBN 978-952-60-7391-0 (pdf)

Thousands of pages have been written on principles of road lighting in the 122 years of electric lighting research. In 1913, the CIE (Commission Internationale de L'eclairage, i.e. International Commission on Illumination) was founded in Berlin to study the photometric properties first of gas lighting, research that later expanded to other light sources such as incandescent lamps. Much ink has been expended on technical argumentation about such lighting since then, enough to create a legacy of technical road and street lighting, thanks to persistent efforts by technically oriented people. In 1931, the CIE introduced an international trichromatic colorimetry and photometry system, known as the CIE System. Engineers became active in illumination engineering societies.

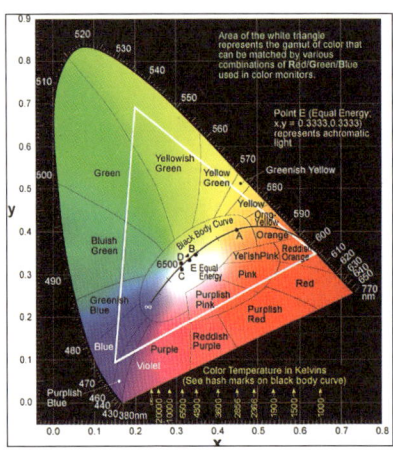

Figure 91: "CIE system." Photo: Author's dissertation ISBN 978-952-60-7391-0 (pdf)

It's useful today to study road lighting with open eyes and not judge it by earlier activities, solutions, norms or recommendations. In our rapidly changing world, road and street lighting applications need new thinking, new ideas and new collaborators. Modern lighting design requires an open mind unfettered by earlier judgments or recommendations. Collaboration among road and street lighting experts, vehicular lighting experts, architects, and biologists ought to be the primary focal point. In 1931, scientists had no idea that the LED would be the only light source in road lighting in 2022, just as we may have only a dim concept of how road lighting might become unnecessary after 2030.

Road lighting today

Roadway lighting is meant to improve traffic safety, assist in efficient traffic movement, and facilitate use during darkness and in all kinds of weather. As a supplement to vehicular headlights, fixed lighting can enable the motorist to see details more distinctly, locate them with greater certainty, and react safely to roadway and traffic conditions. Over almost a century of strong technical development in road lighting, the arguments for it have been fully developed through experiment. Road lighting issues are classified in various ways: by road and street type, amount of traffic, road surface, lamps, luminaires, measurements, etc.

There are hundreds of tables and figures available to describe road, street and light-traffic lighting classifications and recommendations for each. Here are just a few examples:

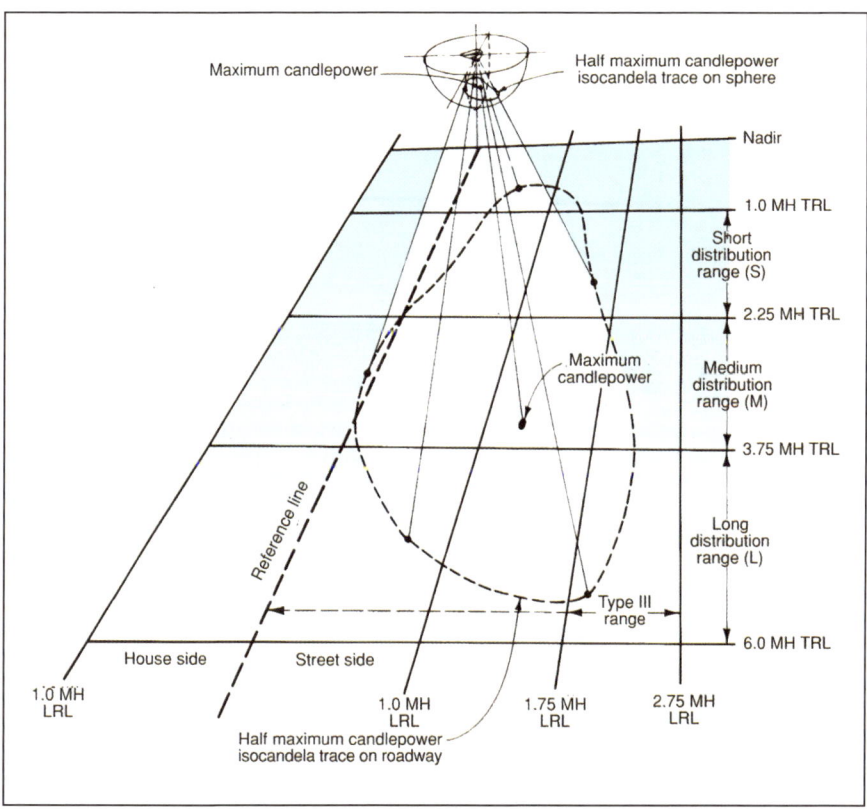

Figure 92: Diagram showing projection of maximum candlepower and half-maximum candlepower isocandela trace from a luminaire having a Type III – Medium distribution on the imaginary sphere and the roadway. Sinusoidal web and rectangular web representations of the sphere are also shown with maximum candlepower and half-maximum candlepower isocandela trace. Photo: "The IESNA Lighting Handbook, Reference & Application," Ninth Edition.

Electrical Info @facebook.com/elarchisocial www.ELarchi.com

CIE 115 - Road lighting norms

Lighting class for motorised traffic

Description of Road		Light Class
high speed roads with separate carriage ways. eg: motorway, express-road	traffic density and road complexity: - high - medium - low	M1 M2 M3
high speed roads, dual carriage way roads.	traffic control, separation and mix: - poor - good	M1 M2
important urban traffic route. eg: radial and distributor roads	- poor - good	M2 M3
connecting less important road. local distributor roads, residential roads.	- poor - good	M4 M5

Lighting Class	Dry L_{av}	Dry U_o	Dry U_l	Wet U_o	TI in %	SR
M1	2	0.4	0.7	0.15	10	0.5
M2	1.5	0.4	0.7	0.15	10	0.5
M3	1	0.4	0.6	0.15	10	0.5
M4	0.75	0.4	0.6	0.15	15	0.5
M5	0.5	0.35	0.4	0.15	15	0.5
M6	0.3	0.35	0.4	0.15	20	0.5

L_{av}	Average Luminance in cd/m2
U_o	Overall uniformity
U_l	Longitudinal Uniformity
TI	Threshold increment (represents glare)
SR	Surround Ratio

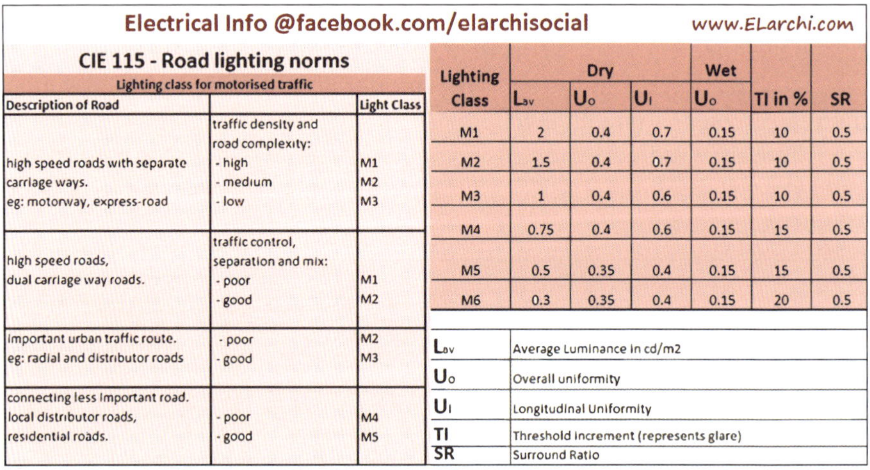

Figure 93: CIE 115 Road lighting norms. Photo: Pinterest.

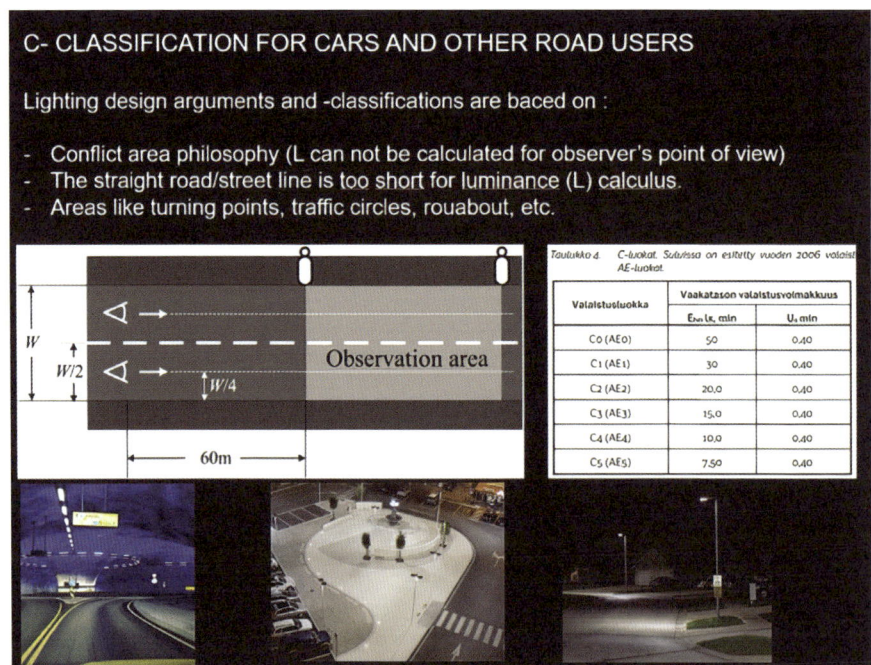

Figure 94: Photo from author's slide show lecture at USN University, Norway, concerning Road Lighting technical treatment.

Impact of current road lighting on biodiversity

Globally, millions of kilometers of road and street lighting networks are the main producers of light pollution in outdoor lighting. Dazzling road lights attract insects with devastating consequences. Light pollution directly to the sky and the reflections back from the sky into the vegetation of the roadside areas disturb fauna in many ways, and we humans also lose the night. The great problem we face in trying to solve this problem is a frivolous attitude, a reluctance to know, a lack of knowledge, a complete lack of understanding, and eventually a non-belief in the devastating effect of light pollution on fauna.

Insects

A *Suomen luonto* (Finnish Nature) magazine article in 2020 asks, "Are insects disappearing, and how can this be studied?"[16] According to research in 2017, the biomass of flying insects in Germany had decreased by three-quarters in just 27 years; in Holland, 84 percent in 130 years. In Great Britain, a third of the 353 bee and flower fly species have experienced a reduction in habitat in 33 years. It has been found that bumblebees have decreased by almost one-fifth in Europe and by as much as 46 percent in North America, compared between the years 1901-1974 and the beginning of the 2000s. This is especially alarming because insects pollinate most of our food, participate in vital life cycles by disposing of dung and carrion, manage biological control, and produce raw materials such as silk.

The best way to study insect extinction globally is through long-term monitoring and modern technologies, a good example being the computer chips implanted on the backs of bees by scientists studying their flight paths and behavior to understand why their species are disappearing.

Illumination of the nocturnal landscape has increased rapidly over the past few decades and is considered an important ecological threat. One of the known impacts of light pollution is the attraction of insects to road and street lights. A research report titled "Long-Term Comparison of Attraction of Flying Insects to Streetlights after the Transition from Traditional Light Sources to Light-Emitted Diodes in Urban and Peri-Urban Settings,"[17] involved collecting insects weekly around streetlights from the end of June until early October in 2011 and 2013. For the nocturnal

---

16 https://suomenluonto.fi/kampanjat/anna-hyvan-mielen-joululahja-tilaa-suomen-luonto/
17 Roy H.A.van Grunsven, Julia Becker, Stephanie Peter, Stefan Heller, and Franz Hölker, 2019.

samples a total of 6718 flying insects were caught over 545 sampling events. That's a rate of roughly 4.5 times the number collected in daytime samples (255 insects collected over 92 sampling events).

Table 1. Number of sample-events per location per lamp type for the different locations and lamp types (light-emitting diodes -- LED, mercury vapor -- MD).

Location	Light Source	Year	Sample-Events
Urban (Berlin city center)			
Leibnizstraße north	MV	2011	44
Leibnizstraße south	MV	2011	44
Leibnizstraße north	MV	2012	56
Leibnizstraße south	LED	2012	56
Peri-urban (Schulzendorf)			
Jahnstraße	MV	2011	112
Jahnstraße	MV	2012	60
Jahnstraße	LED	2012	60
Brandenburger Straße	LED	2012	60
Brandenburger Straße	LED	2013	39
Helgolandstraße	LED	2013	33
Jahnstraße	LED	2013	73

Table 2. Light sources, mean illuminance at ground level, color temperature of the light, and height of the lamp above the ground.

Location	Light Source	Mean Illuminance (lx)	Mean Correlated Color Temperature (k)	Mean Height (m)
Urban (city center Berlin)				
Leibnizstraße north	MV	9.6	4290	10.0
Leibnizstraße south	MV	9.1	3700	10.2
Leibnizstraße south	LED	9.5	3650	10.2
Peri-urban (Schulzendorf)				
Jahnstraße	MV	10.2	4650	4.1
Jahnstraße	LED	18.6	4036	4.1
Brandenburger Straße	LED	17.5	3360	4.9
Helgolandstraße	LED	7.2	2990	4.4

Figure 95: The article goes on to explain the deleterious effects of artificial light, which extend beyond attracting insects to light sources. Artificial light can affect immune responses, predator avoidance, and interactions with plants. Moths reduce night activity when illuminated. This has been shown to affect pheromone production, mating and feeding behavior.

Figure 96: "Street light at night" by dbgg1979 is licensed under CC BY 2.0.

Aquatic insects

Frontiers in Environmental Science published a great 14-page study in 2017[18] that presents research results from a large-scale field experiment on the effect of light pollution produced by street lighting on the aquatic-terrestrial linkage.

---

18 "Artificial light at night affects organism flux across ecosystem boundaries and drives community structure in the recipient ecosystem" Alessandro Manfrin, Gabriel Singer, Stefano Larsen, Nadine Weib, Roy H.A. van Grunsven, Nna-Sophie Weis, Stefanie Wohlfahrt, Michael T. Monaghan and Franz Hölker

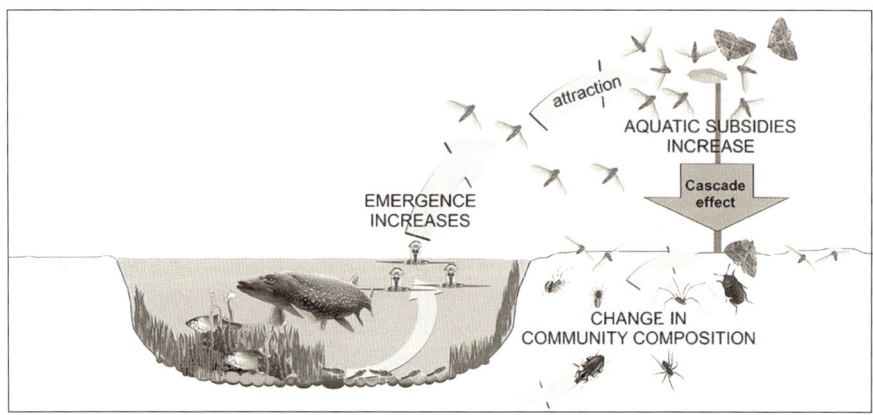

Figure 97: Conceptual figure depicting how artificial light at night (ALAN) increases the movement of aquatic insects into the riparian area because the light increases emergence and attracts aquatic insects. The community of ground-dwelling predators and scavengers on the water's edge is altered in the lighted area, affecting behavior. Some normally night-active riparian spiders, for example, extend their activity into the day. Changes in normal behavior are probably a result of the increase in prey derived from the water. According to researchers, ALAN (Artificial Light At Night) changes the flow between healthy ecosystems on both a regional and global scale.

Figure 98: CCLEX Bridge night view, Cebu City. Photo: Patrickrogue 01

The effects of light pollution on ecosystems can be minimized and possibly even eliminated, but this calls for cooperation among various professional groups: darkness (lighting) designers, biologists, ecologists, engineers, landscape designers, etc. Photo from: "Artificial light at night affects organism flux across ecosystem boundaries and drives community structure in the recipient ecosystem."

Birds

As found in a 2019 review,[19] the use of artificial light at night (ALAN) is increasing exponentially worldwide, including street lighting projects, accelerated by the transition to new efficient lighting technologies (LED units). However, ALAN and the resulting light pollution can cause unintended physiological consequences. In birds (and all other vertebrates), production of melatonin—the "hormone of darkness" and a key player in circadian regulation and a balanced and happy life—is suppressed by light pollution.

The focus in this book is on birds, as light pollution (direct and indirect) from road and street lighting penetrates deep into nature in areas near roads, where birds often nest. Birds are also attracted by bright lighting, and LEDs are exceptionally bright because the light distribution surface is so small while the light output is similar to that of older and larger lighting units with high pressure sodium (HPS) and metal halide (MH) lamps.

As cycles of natural light and darkness are one of the most predictable cues in the environment, light is the dominant environmental synchronizer in most organisms, including vertebrates. Most organisms evolved to use changes in irradiance, spectral composition, and direction of light during dawn and dusk provide a reliable indicator of the time of day. We humans extend our activity and productivity into the night, thanks to ALAN. However, ALAN results in light pollution that disrupts the natural light and dark cycles for other creatures.

---

19 "Light Pollution, Circadian Photoreception, and Melatonin in Vertebrates" Franz Hölker, et al.

Figure 99: Light pollution penetrates into bird living areas. Photo by Robert Natkay, licensed under CC BY SA 4.0.

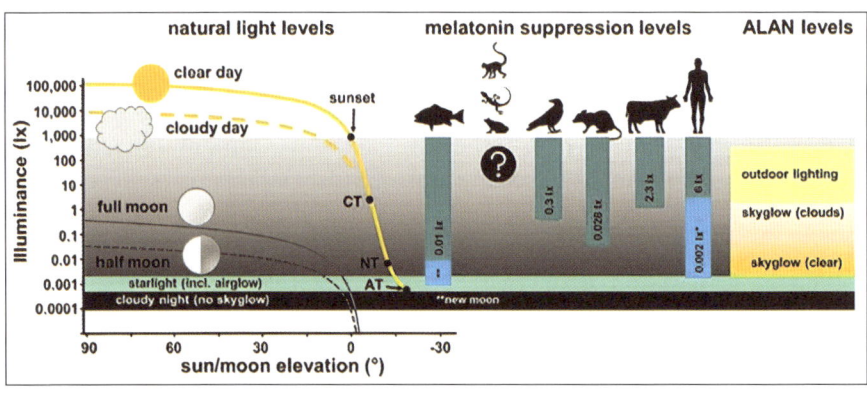

Figure: 100 From the review titled "Light Pollution, Circadian Photoreception, and Melatonin in Vertebrates,"[20] this graphic presents the minimum light levels needed to suppress melatonin in vertebrate groups relative to natural and artificial light sources. Left panel: illuminance during day, twilight, and night as a function of the angle of elevation of the sun and moon; the shaded gray bar

---

20 Franz Hölker, et al

represents twilight (CT—end of civil twilight, NT—end of nautical twilight, AT—end of astronomical twilight), the light green bar represents illuminance in a clear sky, while the black bar represents illuminance during an overcast night; the yellow solid line is sun illuminance on a clear day, while the yellow dashed line is sun illuminance on a cloudy day. The gray solid line is moonlight at the full moon, while the gray dashed line is moonlight at the first and last quarter. Central panel: minimum melatonin suppression levels published in the studies reviewed for various vertebrate groups: fish, amphibians, reptiles, birds, rodents, ungulates, non-human primates, and humans. Right panel: Typical illuminance levels from ALAN; * indicates the minimum level of monochromatic light (460 nm) that, in controlled laboratory conditions, suppressed melatonin. This also indicates there is naturally a significant suppression of melatonin at the time of the new moon.

Several conclusions can be drawn from the results of this review of the effects of light pollution.

First, ALAN exposure at night suppresses or markedly reduces melatonin production and release in birds, even at an illuminance of just 0.3 lx. This change is usually accompanied by changes in behavior, especially shifts in activity patterns. Birds exposed to artificial light are more active at night and in some cases more active overall on a daily basis. Second, as only three studies were specifically designed to test the effects of ALAN on melatonin in birds, more studies, and on more species, are needed to increase our knowledge in this field. Studies should also pay attention to the choice of light bulbs used, as well as the experimental setup. Most studies used wide-spectrum light sources, so there's a gap in research on the effect of various ALAN wavelengths on melatonin in birds.

Finally, we still don't know much about the wider implications of night-time melatonin suppression due to ALAN in birds. What do strong shifts in the timing of activity because of light pollution mean for the health and fitness of birds? This may have significant implications for predator–prey relationships and energetics. So future studies should consider measuring melatonin rhythms under ALAN in wild birds. This question is equally important for domesticated species like chickens, where animal welfare issues also need to be considered.

Millions of road- and street-lighting fixtures produce both direct and indirect light pollution that penetrates deep into the natural life along roadsides, where birds and many other animals live. The problem is exacerbated by the extreme brightness of the light distribution surfaces of the new LED lights. These are increasingly more efficient at producing light than the old high-pressure, metal halide and mercury vapour lamps, so birds are not only naturally drawn closer to roadsides, the strength of the light they are subjected to easily exceeds that needed to suppress melatonin secretion. The numerical value of the strength of light produced by LED lights on minor roads is as much as 45 times that of the old lights.

Being exposed to nocturnal light significantly suppresses or reduces secretion of the nighttime hormone melatonin at a strength of just 0.3 lx. Even this strength of light, apparently insignificant in the modern world, causes strategic changes in avian behavior, making birds more active at night than they would otherwise be.

Only a few studies have been published describing these effects on bird behavior, and the wider effects of melatonin suppression are unknown. Future research will undoubtedly focus on replicating these findings in various species and numbers of birds and on the effects of different kinds of lights and wavelengths on melatonin secretion. We are aware of the optical sensitivity of birds and the spectra of LED lamps used in street lighting, which facilitates research. Further research could challenge the wider effects of this lighting on biodiversity, for example with regard to how it affects predator-prey behavior. Given that, according to the WWF, we have lost 70 percent of our wild birds in the period 1970-2020, it's of utmost importance to conduct research measuring melatonin rhythms in wild birds in areas with different types of electrical lighting.

Daniel Klem, Jr.[21] recalls walking in urban areas at 3 a.m., "when to us humans our vision is just about nonexistent, except when aided by lighting. But birds were very active at 3 a.m., and lethal strikes on windows were common."

Although birds naturally begin to sleep at twilight, at least some passerine birds move actively even during astronomical twilight, when illuminance levels are 0.001lx. Road and street lighting turn on automatically between sunset and civil twilight, when bird melatonin suppression has already begun (illuminance level over 0.3 lx). Interior lights as façade lighting are also on but don't have a strong influence on melatonin suppression if used according to the curve below. (Low interior lighting level eliminates the mirror effect from the bird's perspective but is not at the level of attraction.) More pragmatic research needs to be done.

As road lighting usually remains on overnight, light pollution penetration into the vegetation is a major problem for birds if their nesting area is within the area affected by the light pollution and not deeper in the forest. Again, more research needs to be done in this field.

---

21 Ph.D., D.Sc., Professor of Biology, and Sarkis Acopian Professor of Ornithology and Conservation Biology Muhlenberg College, Allentown, Pennsylvania, USA

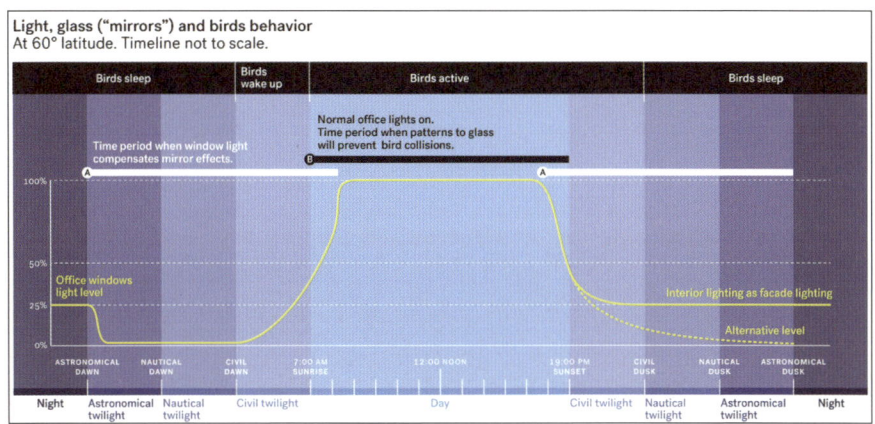

Figure 101: Hewing to the yellow light curve as a façade lighting solution does not interfere with the secretion of melatonin by birds. The window acts as a lighting unit that illuminates the surroundings of the building. Well-designed building lighting provides "whispering light," one of the principles of a nocturnal Darkness Design master plan. (Please note that the yellow curve does not represent the vertical lighting value of the window, but the horizontal value of the interior lighting on the floor, of which only a modest part reaches the window and passes through it.)

Road lighting, clouds, and light pollution

Many elements in road lighting, according to research, interfere with efforts to conserve biodiversity. They can be considered and eliminated in new nocturnal design guidelines due to the almost endless design and implementation possibilities of road and street lighting solutions, in tandem with advanced intelligent headlight solutions for vehicle lights.

Sample road lighting calculation

Computer calculation of the effect of cloud reflection caused by road lighting on area where birds may experience suppression of melatonin.

Figure 102: Factors in the calculation: Illuminated road area (30m x 500m), road surface reflectance R3, selected luminance 1.61 cd/m2, LED luminaires. Clouds are stratus clouds, height from the ground 500m. Photo: Pauliina Oksanen M.Sc. (Eng.)

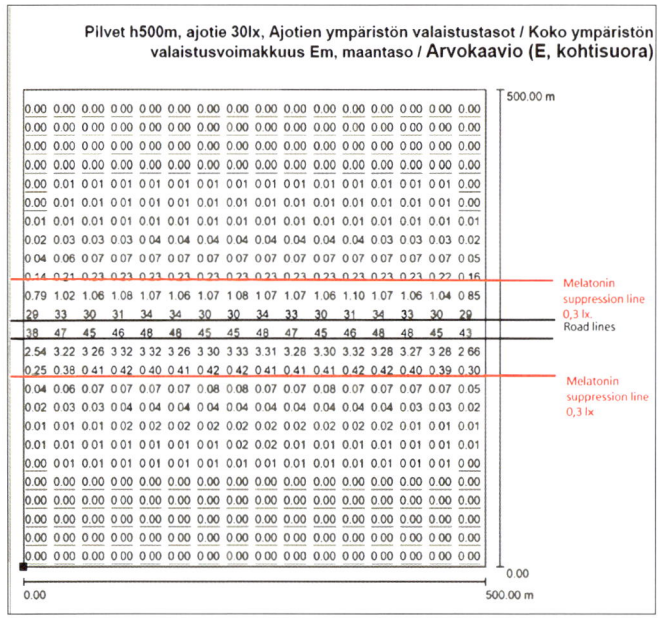

Figure 103: Light values that may cause melatonin suppression in birds extend from the edge of the road lighting toward the forest to a depth of 40 meters.

(This means there is a disturbance area of 40,000 square meters of bird habitat.) Calculation: Pauliina Oksanen M.Sc. (Eng.)

Stratus clouds are low-level clouds characterized by horizontal layering with a uniform base, as opposed to convective or cumuliform clouds, which are formed by rising thermals. More specifically, the term *stratus* is used to describe flat, hazy, featureless clouds at low altitudes, varying in color from dark gray to nearly white. The word *stratus* comes from the Latin prefix *strato*, meaning "layer." Stratus clouds may produce a light drizzle or a small amount of snow. These clouds are essentially above-ground fog formed either through the lifting of morning fog or through cold air moving at low altitudes over a region. Some call these clouds "high fog" for their fog-like form. While light rain may fall, this cloud does not indicate much precipitation. The height of stratus clouds varies from 0 to 2400 m. In the calculation, stratus clouds were simulated on a glass plate at a height of 500 m with a mirror reflection value of 30%.

Figure 104: Stratus clouds. Photo: Daniel Akuoko, licensed under CC 4.0 International

Modern vehicle light systems + intelligent LED road lighting

Problem 1: The halogen headlight era

As quoted above, "Roadway lighting can improve traffic safety, achieve efficient traffic movement, and promote the general use of the facility during darkness and under a wide variety of weather conditions. As a supplement to vehicular headlight illumination, fixed lighting can enable the motorist to see details more distinctly, locate them with greater certainty, and react safely to roadway and traffic conditions present on or near the roadway facility." The text defines car headlights as the primary light source and fixed road lights as the *complementary* light. At the time this was written, roadway lighting classifications and recommendations did not take into account halogen headlights in the calculations of what level is necessary.

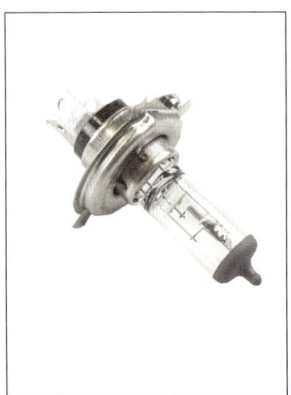
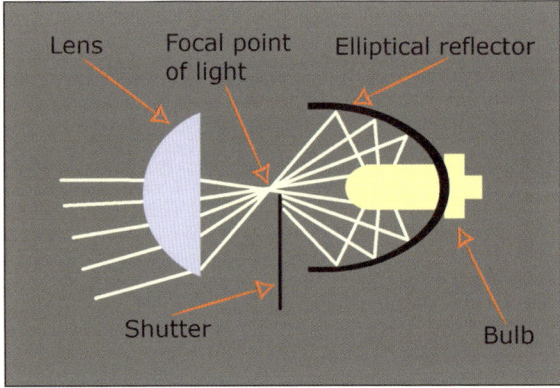

Figures 105 left and right: The simplest type of halogen headlight is the reflector type. The headlight assembly consists of a bulb situated at the focal point of the reflector. When the bulb is turned on, the light spreads all over the shiny surface of the headlight and, because of reflection, the light beam is directed towards the road. Photo: GoMechanic

Headlighting has always presented a dilemma for the lighting designer. On one hand, the luminous intensity directed down the road is very important to the driver for maintaining lane control, detecting roadway obstacles, and giving the driver a feeling of wellbeing behind the wheel. On the other hand, glare meeting oncoming drivers and reflecting in the rear-view mirror of a car ahead restricts how much light can be safely delivered to the roadway.

So headlight beam distribution has always been an optimization between forward visibility and minimized glare.

Problem 1 solved: Intelligent LED headlights + controlled LED road lights

We are living in a new era of LED lighting in the automotive world. A fast transition is under way to LED headlights, just there has been a swift change to LED and electronic control in road and street lighting. Taking that one step further, intelligent LED headlights may make it possible to replace road and street lights, while saving energy and preserving biodiversity otherwise threatened by excessive ALAN.

Car manufacturers have already developed and are marketing such intelligent LED headlights. The possibilities are showcased in a video that can be viewed at https://www.youtube.com/watch?v=Zur2KcBOUf4, which describes the complex headlight solutions of the car manufacturers Mercedes Benz, Audi, and BMW: Mercedes-Benz's NZ MultiBeam LED vs. Audi Matrix LED vs. BMW Intelligent Headlight Technology 5.1.2017.

Mercedes Benz as an example

In the new E-Class, the optional high-resolution Multibeam LED headlamps, each with 84 individually controlled high-performance LEDs, automatically illuminate the road with a hitherto unsurpassed, precision-controlled distribution of exceptionally bright light—without dazzling other road users. That's because this grid allows light distribution in the left and right headlamps to be controlled separately and adapted to the changing situation on the road, quickly and dynamically. All functions of the Intelligent Light System in low-beam and high-beam mode can furthermore be directed for the first time purely digitally and without mechanical actuators, including, as a world first, a purely electronically implemented active light function.

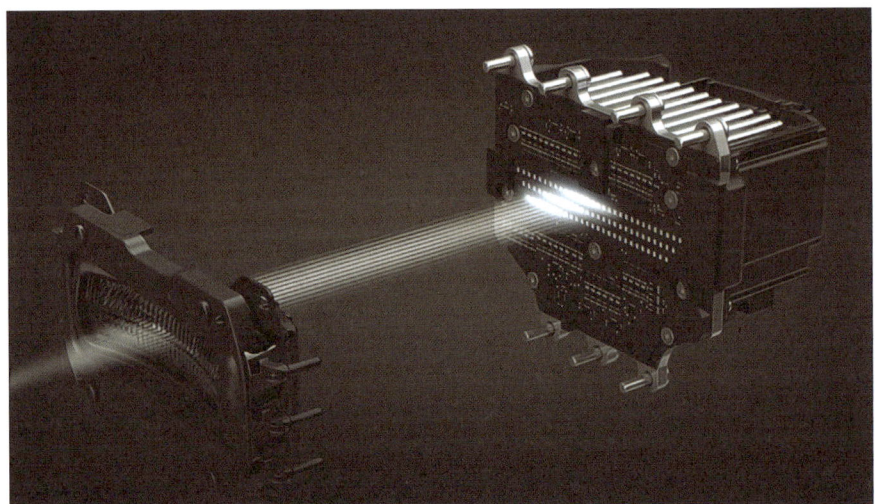

Figure 106: Multibeam LED headlights, Three-stage Precision Optical System. Photos: Mercedes-Benz E-Class: Multibeam LED

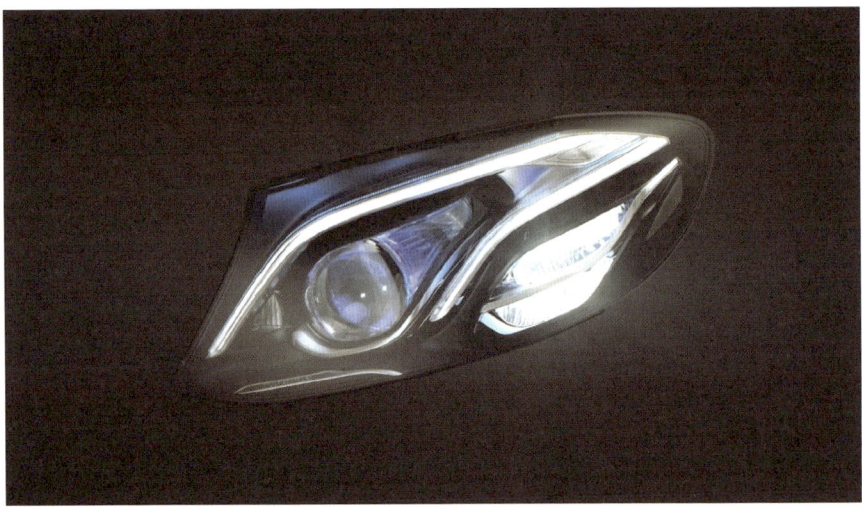

Figure 107: Multibeam LED headlights, Three-stage Precision Optical System. Photos: Mercedes-Benz E-Class: Multibeam LED

Figure 108: For perfect visibility, the adaptive Multibeam LED headlights with individually controlled LED lights respond to the traffic situation. Photo: Mercedes-Benz E-class: Multibeam LED

Tests under various conditions

Intelligent headlight technology is evolving at a tremendous pace. Headlight solutions from car manufacturers are being researched and tested internationally by several players. From these, information and feedback are passed on to manufacturers, who take advantage of valuable feedback. For example, in Finland, valuable research and testing is being done by master's student and headlight expert Jari Pitkäjärvi, whose test results in Finland's often extreme driving conditions are appreciated and utilized by international car manufacturers. Pitkäjärvi is commissioned by the Finnish technology magazine *TM*.

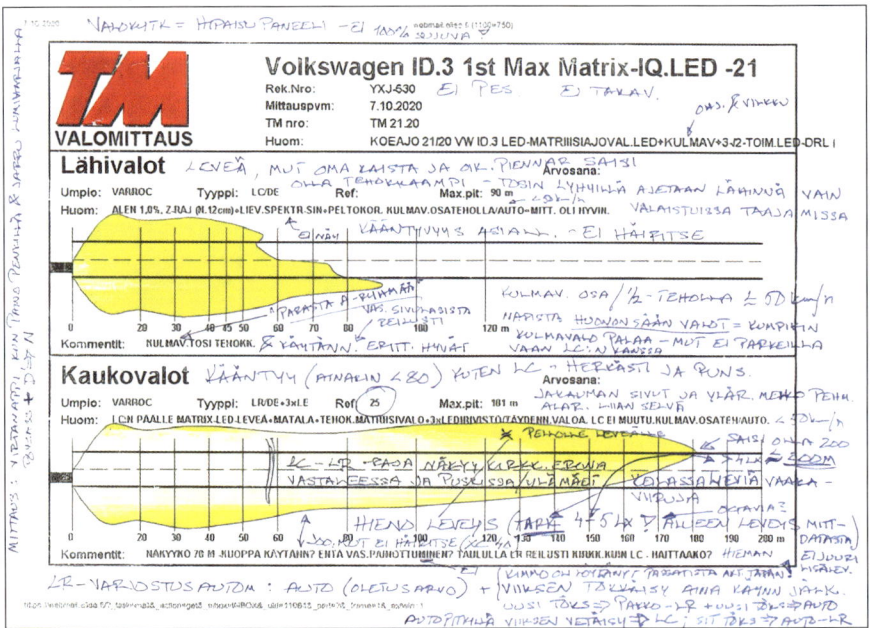

Figure 109: A test result showing how a headlight system works in real life, by Finnish researcher Jari Pitkäjärvi, who tests vehicle lighting systems for *TM* (World of Technology) magazine. New intelligent headlights for cars are being researched and tested by various parties, which typically give developmental feedback to the manufacturers. This result is for the VW ID.3 first edition[t] Max Matrix-IQ.LED – 21. Photo: Jari Pitkäjärvi

Problem 2: Fixed-value road lighting

As a supplement to vehicular headlight illumination, fixed lighting can enable the motorist to see details more distinctly, locate them with greater certainly, and react safely to roadway and traffic conditions present on or near the road. In practice, this means a fixed choice of lighting value for road lights, in accordance with the lighting recommendations of each country.

Figure 110: Left: High-pressure sodium lamp fixture. Right: Philips SGS luminaire for Tubular High Pressure Sodium Lamp. No dimming option; a solution just covering the road lighting recommendations. Road lighting remains on continuously, regardless of traffic volumes and time. Photos: JO gallery. Photo Left: Robert Natkay, licensed under CC BY SA 4.0

Problem 2 solved: Electronically controlled LED road lights

The rapid development of electronics and wireless lighting systems in the LED world has enabled road and street lighting to be configured any way a customer wants. In the author's personal design experience, all desired lighting functions are now possible in all lighting sectors.

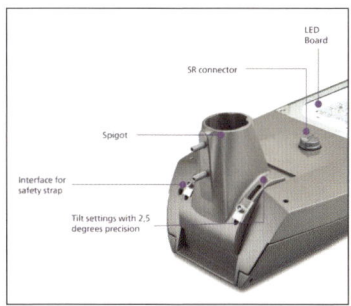 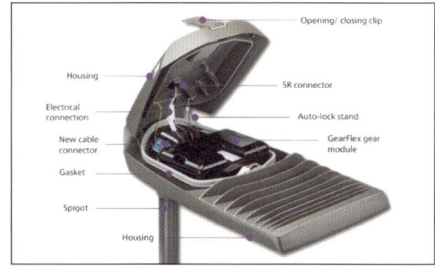

Figures 111 above and below: Philips Luma gen2 LED Road and urban luminaire. Dimming: Stand-alone Dyna Dimming, DALI, Lunanote, Mains dimming. Connectivity options: Interact City, Easy Air, Watstopper motion sensor. Photo: Philips Luma gen 2 brochure.

Summary: Road, Street, and Vehicle Lighting

It's high time we overcame our resistance and started encouraging close and comprehensive cooperation among biologists, road and street lighting experts, and vehicle lighting experts. We'll need a clear pattern of collaboration that includes interest groups of experts from different disciplines and successful companies in the respective fields. The work should be led by an international organization, such as the UNEP (United Nations Environment Program). The aim of the working group should be to minimize and partially eliminate road and street lights as an unnecessary complement to intelligent vehicle headlights.

In addition to protecting biodiversity, this offers huge potential for energy savings. LED headlights alone, common already even without intelligent operation, would easily reduce the need for road lighting by 20%. This would offset an annual increase in light pollution of 2.2% for almost ten years.

Once more we'll quote world-renowned lighting expert John D. Bullough, who was impressed by the robust performance of ADB headlights under evaluation by researchers. "Our evaluation revealed that the system can offer tangible safety benefits: high-beam forward visibility with low-beam glare. Indeed, ADB systems work so reliably that there may be little reason to limit them to conventional high-beam intensities, since they can rapidly reduce intensity (and glare) that would be observed by others."

Bullough says an advantage is that drivers could use their high beams more frequently, rather than only 25% of the time, as is typical today even on dark rural roads. "This could have implications for road lighting, since standards state that reasons to install lighting include reducing headlight glare and supporting vision beyond headlights' reach. To the extent they sustain driver visibility without increasing glare, ADB headlights might permit road lighting to wither, at least in certain situations and locations."

More research and demonstration will be needed, he writes, "But less dependence on road lighting could have other benefits, like less conflict between road and vehicle lighting, reduced energy use, fewer poles, and improved connections to our nighttime sky. Are they worth the effort?" To that, I respond: "Yes, they are!"

ADVERTISING LIGHTING

Billboards

Billboards placed in roadside areas should be avoided, for several reasons:
1) When a vehicle is moving at 100 km/h, for example, it advances 28 meters per second. Large illuminated roadside billboards against a dark background distract the driver. In the several seconds it takes to view it, the vehicle may have already traveled more than 100 meters forward.
2) Billboards are big and ugly elements that have no place in the landscape architecture.
3) Illuminated billboards invite birds and insects and disturb their circadian rhythm.
4) Big billboards cause a lot of light pollution and illuminate space for nothing.
5) The commercial benefits of large and illuminated billboards cannot be demonstrated. The author's personal survey showed that the people targeted didn't even remember what the billboard said.

The author believes billboard advertising is close to subliminal advertising, where the stimulus of the message is given so weakly (in this case, almost at a glance) that no conscious perception is made, though the subconscious may sense the message. Subliminal advertising is a difficult entity to manage, which even the European Parliament has studied and considered prohibiting based on surveys from various countries. (For example "*Parliamentary question - E-003522/2015*")

Figures 112: "Bunny," by In Memoriam: saschapohflepp, is licensed under CC BY 2.0.

Other advertisement lighting solutions

Illuminated advertisements have a place in modern society. One of the strangest examples of this is Times Square, New York City. People flock from all over the world to experience its unique madness, while locals avoid it like the plague. It represents classic binaries in life: love and hate, salt and sugar, life and death, city center and rural landscape. It's difficult to find landscape architecture in this place. The brightness of the advertisements ought to be dimmed, while still maintaining the enjoyable craziness of the space and the pulse of life. The light pollution produced to the sky by illuminated advertisements could be minimized by tilting the screens towards the ground some degrees (the angle depending on the structure of the space).

Figure 113: Today, Times Square heaves with tourists as one of the most famous and widely visited locations in the world. Photo: "times square at night," by JOnasIsMyMiddleName:), is licensed under CC BY 2.0.

ULOR-CT (UPWARD LIGHT OUTPUT RATIO – CITY TOTAL)

ULR and ULOR

The beginning of the 2000s witnessed the development of ULR (Upward Light Ratio) methodology, which aims to minimize light pollution from electric lighting directed to the sky. The international organization CIE identified four types of environments for which disruptive light values from electric lighting were categorized. (E1 = Nature-like/Dark environment; E2 = Countryside/little environment lighting, E3 = Suburban area/average environment lighting, E4 = Town city/strong environment lighting). The elements of disruptive lighting were categorized as "Illumination of the sky," "Light to windows," "Light intensity of a light source," and "Luminance of the building." The technical values are presented in the tables below. International manufacturers of lighting equipment immediately began to include in their production lines LED lights that send zero light to the sky: ULOR (Upward Light Output Ratio) = 0.

These recommendations on disruptive light were well-meant and have been in use for over ten years, but they have not led to satisfactory results in practice. According to values obtained from the Suomi satellite orbiting the Earth, light pollution has continued to grow by 2 percent to 3 percent every year, and this is accelerating.

One example of using the ULOR concept in an architectural lighting solution comes from 2008: the Brando luminaire. The "Brando" name was inspired while I was watching the movie *The Godfather* alone one night on the campus of Oregon State University. Seeing Marlon Brando's charismatic shoulders, I knew that this was the perfect character after which to name a new outdoor lighting luminaire model. I dreamed of this luminaire becoming the new giant of the lighting world, with all the charms of Marlon Brando and the indisputable charisma of Humphrey Bogart. The project in which the Brando luminaire was used is described in appendices to my dissertation: https://aaltodoc.aalto.fi/handle/123456789/27886

Figure 114: "Like a boss. Godfather legacy," by Syed Ikhwan, is licensed under CC BY 2.0.

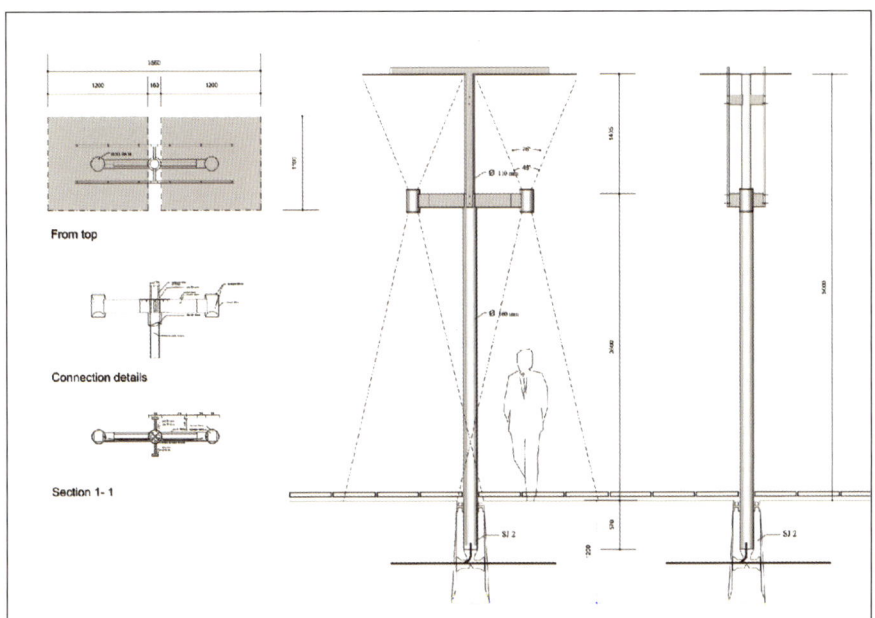

Figure 115: Brando dimensions. Photo: Julle Oksanen's dissertation. Julle Oksanen & Oliver Walter

Light pollution and ULOR

New energy-efficient optical designs are mostly equipped with a new light source, LED, and designed in such a way that ULOR = 0. "Disturbing" light is that part of light from a lighting installation that doesn't serve the purpose for which it was designed.

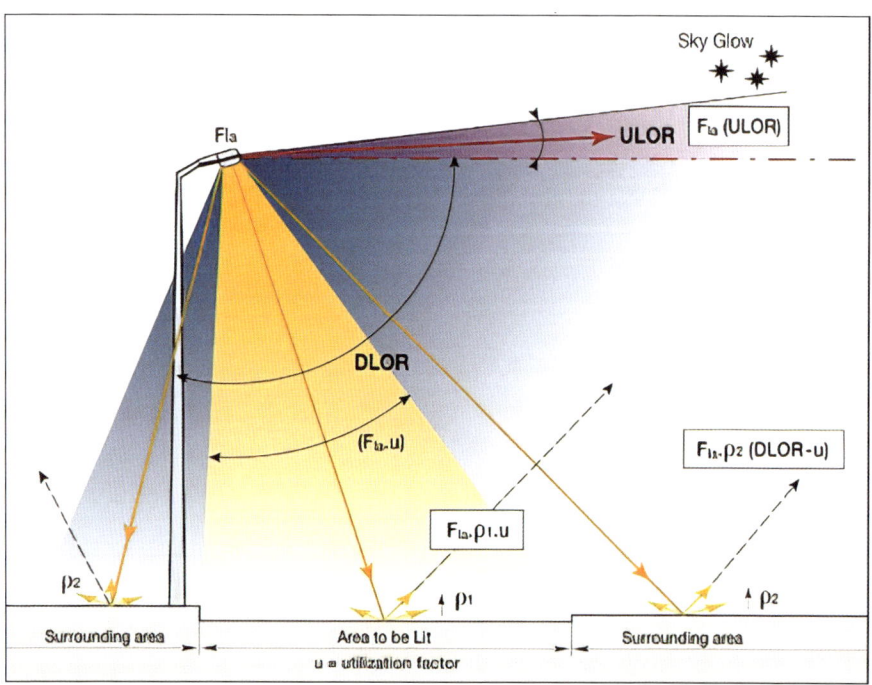

Figure 116: Methodology for outdoor lighting that minimizes sky glow. Photo: CELMA

Contribution of an outdoor lighting installation to sky glow can be determined from data provided by:

- the luminaire manufacturer:

    - Lamp lumen output: F_{la}
    - Upward Light Output Ratio: ULOR [%]
    - Downward Light Output Ratio: DLOR [%]
    - Utilization factor of the lighting installation: u [%]

- contracting authorities and site conditions:

    - Average maintained level of illumination: E [lx]
    - Area of the surface to be lit: S [m²]
    - Reflection factor of the area to be lit: ρ_1 [%]
    - Reflection factor of the surrounding area: ρ_2 [%]

The upward light of an outdoor lighting installation which feeds the sky glow is given by UPF which is made up of the following elements:

- Direct luminous flux emitted upward by the luminaire: $F_{la} * ULOR$
- Luminous flux reflected by the area to be lit: $F_{la} * \rho_1 * u$
- Light reflected by the surrounding area: $F_{la} * \rho_2 (DLOR - u)$

Classification of disturbing light / CIE 150:2003				
Class	E1	E2	E3	E4
Environment	Nature like	Countryside	Suburb area	Town city
Lighting environment	dark	Little envir. Lighting	Average environmental lighting	Strong environmental lighting

Figure 117: Classification of disturbing light / CIE 150:2003

Limit values for the disturbing light

Envir. class	Illumination of the sky	Light of the windows		Light intensity of a light source		Luminance of the building	
		Ev lx		I kcd		L_{av} cd/m^2	L_{max} cd/m^2
	ULR %	evening	night	evening	night	night-time	
E1	0	2	1	2,5	0	0	0
E2	5	5	1	7,5	0,5	5	10
E3	15	10	2	10	1	10	60
E4	25	25	5	25	2,5	25	150

Figure 118: Limit values for the disturbing light

Classifying disturbing light according to various environments, together with its maximum illumination of the sky, ULR % = Upward Lighting Ratio, defines how the luminaire is used. The Brando luminaire is a direct–indirect light producer. The light-producing unit has a single 70W metal halide lamp. Some 50% of light lumens go up on to the white reflector, which reflects light softly around the environment, and the remaining 50% fall directly downwards at a beam angle of 34 degrees, given the rhythm for the space.

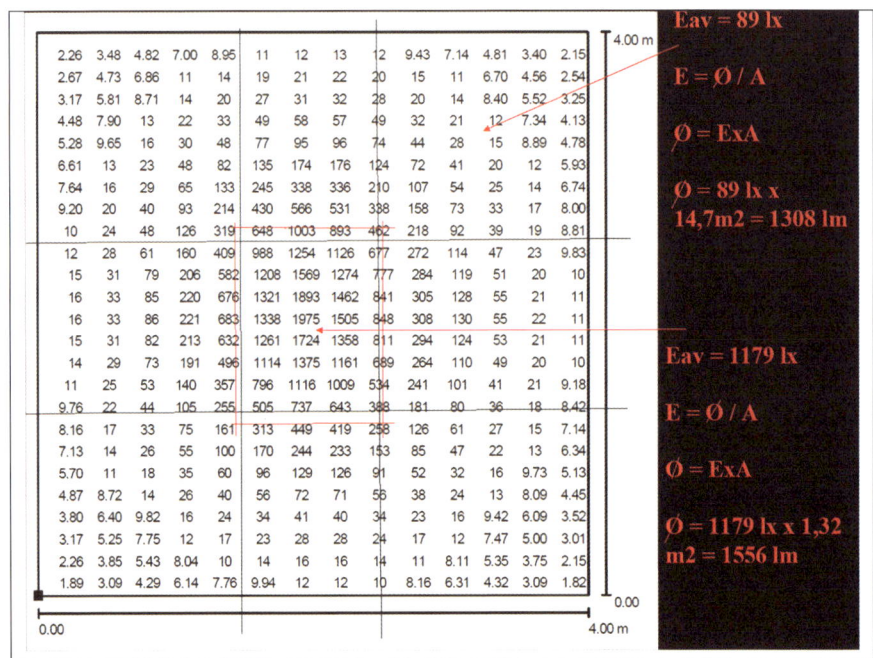

Figure 119: Disturbing light measurements. Photo: Julle Oksanen. In disturbing light measurement, the reflector is marked in red in the middle of the image. Eav on the reflector surface is 1179 lx. When this value is multiplied by the reflection area A (1.3 m2), the total lumen package flowing on the reflector surface is 1556 lm.

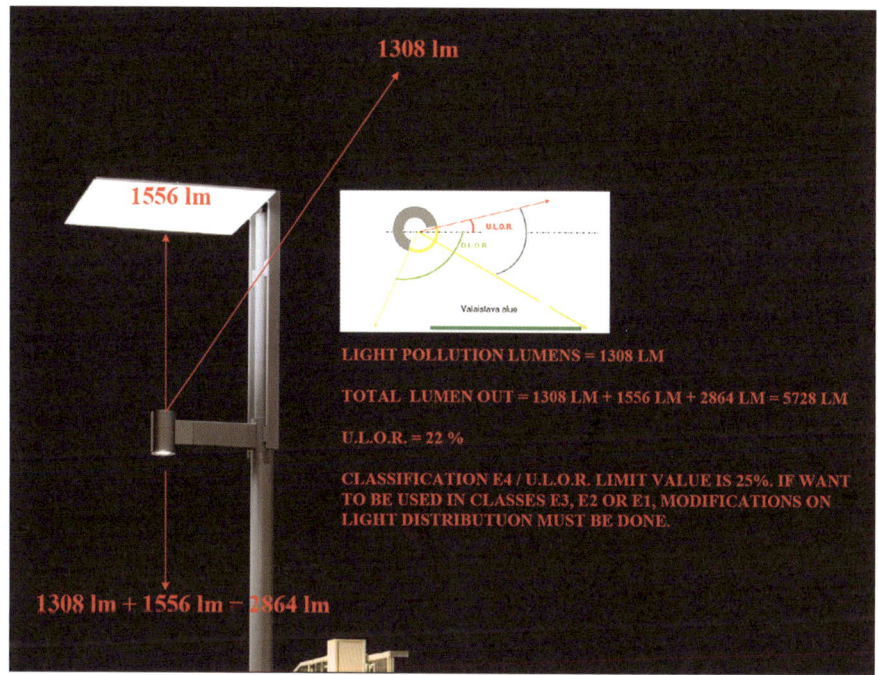

Figure 120: Disturbing light calculation. Photo: Julle Oksanen. The calculation for disturbing lighting (ULOR) gives 1308 lm.

The calculation shows that ULOR = 1308 lm. Total lumen output is 5728 lm. This means that ULOR % = 22%. Thus, the existing Brando can philosophically be used only in town and city areas where ULOR % = 25%.

ULOR-CT

The values of the ULR and ULOR philosophies, which are difficult to design, implement, and measure, can be replaced by a totally new design for eliminating light pollution, repairing old installations, and subsequently measuring implementation with new, modern ULOR-CT methods (Upward Light Output Ratio – City Total).

ULOR-CT differs from ULR (Upward Light Ratio) in that it contains concrete, practical instructions for eliminating the degree of light pollution in a particular area. Using it places appropriate pressure on the design work to eliminate such light pollution. ULR, with its strong theoretical basis, does give designers something to think about, but it doesn't present concrete measures. It only gives theoretical guidelines for light values that are difficult to apply to practical planning for different types of areas—for example, different luminance values for different building facades in different contexts. Design offices typically don't even have a luminance meter to assist with this.

When ULOR-CT philosophy and its new methodology are combined with the nocturnal Darkness Design outlined later in this book, the city space will be significantly freer of light pollution.

ULOR-CT and the Brando luminaire

Figure 121: Desired ULOR % = 0. Photo: Julle Oksanen

The philosophy of nocturnal Darkness Design, meant to protect biodiversity and restore our lost nights, embraces the principle that interior lighting is the best façade lighting. The majority of the light thus reaching into sky/space is reflected from the ground and other city surfaces. This methodology offers an opportunity, through appropriate design of darkness, to provide a harmonious, pleasant nocturnal ambience in the city. It becomes feasible to implement Richard Kelly's hierarchy of light (as described in Chapter 3) on the level needed for a pedestrian to comfortably recognize another person approaching in the dark, as well as R.G. Hopkinson's curve for pleasurable lighting values on the surfaces of various city spaces (also described more specifically in Chapter 3).

ULOR-CT findings can easily be used to design the lighting values of urban structures using a centralized control system like the "Signify" system, to be presented later. The potential energy savings are vast, and the end result an enjoyable and inviting experience of urban space.

It is time to replace the existing ULR (Upward Light Ratio) and ULOR (Upward Light Output Ratio) concepts with a new solution that protects biodiversity, is feasible, and can be controlled in cooperation between energy companies and cities: ULOR-CT (Upward Light Output Ratio - City Total). In this approach, energy utilities and planning officials responsible for urban lighting will need to coordinate to make adjustments to pre-existing problem areas that produce light pollution in space.

ULOR–CT philosophy goes hunting for light pollution at two levels: large-scale and small-scale.

Detecting Large-Scale Light Pollution (global satellite research)

What might be dubbed Large-Scale Light Pollution Detection refers to measuring light pollution caused by lighting on large urban structures that directly faces the sky. A tool that measures such light pollution (mounted on a satellite) orbits the Earth at hundreds of kilometers per hour. What it "sees" proves that the problem of biodiversity-disturbing light is getting worse year by year.

Historically, wherever in the world there has been a rise in gross domestic product (GDP), improvement in the efficiency of light sources has resulted in a tremendous increase in the use of artificial light. Rather than saving energy, better luminous efficacy has resulted in correspondingly greater light use in prosperous parts of the globe. Humans have tended to use as much artificial light as they can buy for about 0.7% of GDP.

Outdoor lighting became commonplace with the introduction of electric light and grew at an estimated rate of 3 percent to 6 percent per year during the second half of the 20th century. As a result, the world has experienced widespread "loss of night," with half of Europe and a quarter of North America experiencing radically modified light-dark cycles.

Figures 122 left and right: A recently calibrated Suomi NPP satellite radiometer designed for night light research 833 km above the Earth's surface shows that between 2012 and 2016, light pollution grew by 2.2% per year and continues to do so. We have lost the night; stars are disappearing from our night views, and energy consumption is increasing. Credits: NASA

Detecting Small-Scale Light Pollution (community level)

Preventing unnecessary light from going to unintended places requires a coordinated effort by professionals. The principle of ULOR (Upward Light Output Ratio), which was created mainly to address exterior fixtures in urban areas, is not sufficient. It needs to be extended to entire residential areas, cities, towns, and villages. This can be achieved by creating a new ULOR-CT (Upward Light Output Ratio–City Total) philosophy and operating model, which I have tested in the city of Turku. As mentioned earlier, this requires utilities and planning officials working together to consider problem areas in their jurisdictions and make adjustments to reduce and preferably eliminate light pollution in space. New lighting plans should be inspected by an ULOR–CT expert.

Conducting a ULOR-CT study

ULOR-CT research is always done with a drone. In its simplest form, this can employ a normal drone that is flown above the selected area to be investigated, with the drone's camera pointing directly to the ground below it. The drone's flight height depends on the area of the section to be measured, selected from the measurement map chosen by the drone user (e.g. city or power plant engineer).

Before the overall mapping of the city's measurement areas begins, it is recommended to make a practice flight, which will facilitate the selection of area sizes.

Wrongly installed floodlights and road lighting installations aimed at the sky are revealed immediately in the images saved to the drone's camera, which can be connected to a laptop or mobile phone.

A more advanced ULOR-CT measurement can be done using a drone with a luminance meter instead of a camera, set up to measure the luminances of the entire area to the sky (e.g. incorrectly installed dazzling lights, too-bright facades, road-lighting hierarchies and levels, etc.).

The more advanced ULOR-CT measurement method requires professional expertise and raises research to a new level. It is worth starting your research with a basic drone, and if you have the knowledge and know-how, you can change later to a more advanced system.

Figures 123 left and right: An ULOR-CT engineer from a power plant launches a professional drone to assess the demarcated areas of the urban structure, with the help of a measurement boundary map. The drone is flown to a reasonable height and the engineer takes a picture of the selected area. Light pollution source points in the image are plotted on the map, and the light point is subsequently re- aimed or replaced with a new lighting unit that meets the ULOR-CT criteria (ULOR = 0). Photo left: principle of the measurement boundary map is not in right scale for normal drone study. An image of author's lecture at Nachtungen, Berlin. Photo right: "Drone", Photo by form PXHere, licensed under CC0.

CHANGING THE PROCESS

In my award-winning dissertation, "Design Concepts in Architectural Outdoor Lighting Design, Based on Metaphors as a Heuristic Tool," I posed four research questions related to altering the focus and methodology of lighting design. These are my summarized responses to each.

1) **Is it possible to create a new architectural outdoor lighting design approach by utilizing heuristics and metaphors as a problem-solving tool in lighting design?**

Although heuristics is not praised in scientific debate, it has been shown to be a pivotal part of both experts' and laymen's problem-solving.[22] In simple terms, a heuristic is a kind of 'mental shortcut' that helps people make decisions and judgments quickly, without spending a tremendous amount of effort on research and analysis. We might use terms like "an educated guess," "intuitive judgment," "common sense" or "rule of thumb" when applying heuristics in areas of work where arriving at an optimal solution is difficult to achieve but an adequate practical solution is possible. It is considered a useful tool to generate and transmit ideas, and to cope with problematic situations.

In the rule-governed "technical lighting" design process, where lighting design decisions are based on rules, norms and technical recommendations, the heuristic method is rarely used. In the architectural lighting design process, on the other hand, the heuristic approach is a very important tool, after the selection of a conceptual design approach, and is involved in creating a metaphor for the design.

Aristotle was one of the first to point out the effective role metaphors can play in the creative process, explaining that a metaphor "consists in giving the name that belongs to something else."[23] The importance of metaphors is that "ordinary words convey only what we know already: it is from metaphor that we best get hold of something fresh." (The author's own example: "My wife is my sun.")

Both heuristic and metaphorical thinking are valuable elements in exploring this research question.

---

22 Dreyfus & Dreyfus 2005; Kahneman et al 1982
23 Picot, J-C. 2006

2) **Can we find feasible design tools to change the direction of a 100-year-old technical lighting design practice towards more artistically oriented, architectural lighting design processes?**

There is at least one scientific "tool" that allows us to utilize most of the valuable light technology findings, with all their scientific terminology, amid the uncertainties that arise when seeking practical solutions in real life architectural lighting design projects. That tool is the "Pragmatic Theory of Truth," which generally holds that a proposition is true if it is useful to believe. A prime example is Isaac Newton's laws of physics. We continue to design buildings today according to these laws, though we now know they are not theoretically perfect. That's because they are still precise enough for practical calculations and work in real life.

3) **In this transition, is it possible to apply those tens of thousands of pages of technical lighting research results we already possess to arrive at aesthetic lighting solutions?**

This book presents some practical project examples of how technical lighting research can be used in creative ways, using heuristics, metaphors, and the Pragmatic Theory of Truth as innovative tools.

4) **What are the benefits of using the first-generation pragmatism of C.S. Peirce and William James to utilize technical research results to serve architectural lighting design?**

Technical lighting research is important for the ongoing development of the industry, but the innovative use of pragmatism immediately gives meaning to those research results, releasing the findings to serve projects in practical and aesthetically pleasing ways.

New ways of thinking about lighting must resolutely strive to eliminate lighting solutions that harm biodiversity, replacing them with solutions that follow nocturnal Dark Design principles as presented in this book. The light pollution element has to be seriously considered.

The new paradigm's technical and organizational elements can be harnessed to protect biodiversity by removing light pollution from architecture, while simultaneously achieving significant energy savings.

Instead of project-specific lighting design teams as we have them now, the ideal team for creating biodiversity-friendly nocturnal design features collaboration between: 1) a darkness designer (i.e., a lighting designer with the right attitude); 2) a lighting engineer; 3) the architect; 4) biologists; and 5) the client's representative.

The next chapter explores how this new paradigm of Darkness Design brings these collaborative skills and cutting-edge technologies together.

CHAPTER 3: PARADIGM FOR DESIGN OF NOCTURNAL SPACES

INTRODUCTION

Lighting Design and Numerical Light Management

Paradigms 1 and 2 in this book addressed elements that have shaped the Light Pollution Paradigm, technical issues in lighting and factors managed according to numbers, and examples of projects by which biodiversity can be protected by eliminating light pollution.

In every country, technical lighting recommendations are based on internationally approved, mathematically set parameters that have been derived from empirical tests. Even the test of whether light is sufficient to offer comfort while approaching another person in the dark is based on numbers: empirical tests have specified the value of the semi-cylindrical illumination strength needed to recognize an approaching face and the minimal distance for pleasant movement.

Lighting design is based on keeping within these values, which can be measured with a simple, universally available device, the lx meter. Virtually all electrical engineers have such a device, as do architects interested in lighting. When the focus is on using numbers to manage light, the goal is only to ensure that the project meets the recommended lighting values, not to ensure the space is pleasurable in terms of how it manages darkness—which is how it ought to be. Management by numbers is a good servant but a bad master.

Darkness Design—The Management of Nocturnal Spaces

Economic and construction considerations have largely determined how urban structures are built in the modern age, frequently neglecting elements that traditionally and naturally make us feel at home in the urban space and create a feeling of civic pride. No wonder we often experience positive sensory experiences when visiting the "Old Town" of a city. If we want to render urban development more human in the future, designers need to cater not only to economic utility but also take into consideration the enticing elements inherent to cities that appeal to our senses.

Making sensitive use of light and correctly designing darkness are important here. We sense light physically and emotionally. In a city center, the vast amount of LED light compressed into the small surface of a road light may actually be experienced as painful, while an oblique whisper of indirect light reflected from the surfaces of urban structures can be greatly enjoyed. Lighting solutions that offer a wide distribution of light, frequently manufactured as "custom made," can be glare-free, imbuing the space with a peaceful essence. During daylight hours, these large lighting elements should blend easily into the architecture.

The constant variation in daylight and the inevitable descent of darkness are enjoyable, rhythmic events in human life. We are particularly enthralled by the rising and setting of the sun. These variations significantly enrich urban spaces such as bridges, statues, water elements, groups of trees and other elements, stimulating the mind. Artificial light in these places needs to be glare-free and should bind the space in holistic harmony. The third paradigm in this book concerns tools to be used for the controlled design of darkness, which can be used to replace existing designs that cause light pollution, or to create something completely new. The arguments in this paradigm can be used to create nocturnal spaces that protect biodiversity, save energy, and create human living rooms in urban spaces that are pleasant and safe. Design models should be governed by our biologically inherent love of twilight, and our need to enjoy it in safety.

The Spirit and Requirements of Darkness Design

Reading what follows can help lay persons enhance their understanding of this complex issue. Among lighting professionals and experts in related fields, absorbing and applying this knowledge will require a more profound mastery of the principles of lighting. The aim of the book is to motivate people with an interest in lighting to become active in eliminating light pollution and its effects on biodiversity within the limits of their own capacity. It is my sincere personal hope that students of related disciplines (e.g. ecology, philosophy, architecture, engineering, spatial art, etc.), when considering their final projects and theses, would consider delving into the wonders of the domination of darkness.

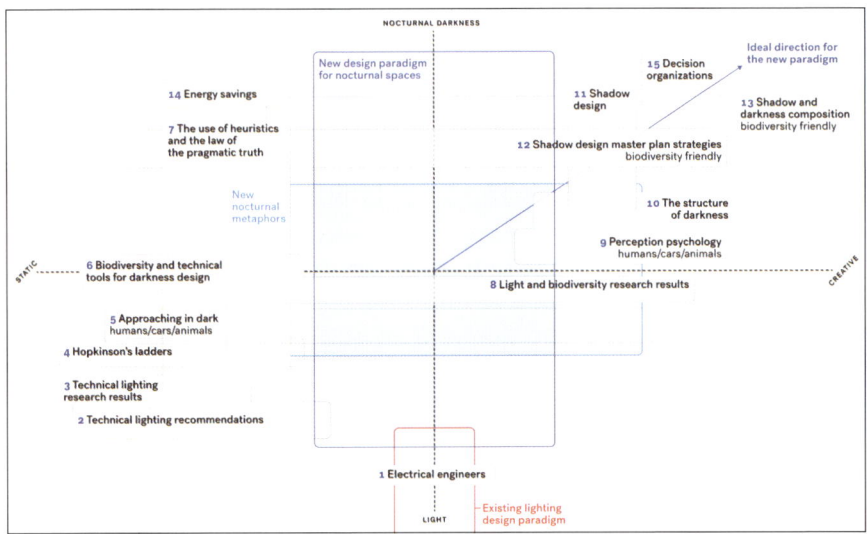

Figure 124: Elements/tools of the Darkness Design Paradigm, drawn on a cross section of Light & Nocturnal Darkness (vertical line) and Static & Creative (horizontal line). Photo: Julle Oksanen and Safa Hovinen.

BIODIVERSITY AND TECHNICAL TOOLS FOR DARKNESS DESIGN

Engineers have accomplished much in terms of increasing the quantity and distribution of light. But the field of lighting design is unbalanced and skewed. Many projects lack aesthetic merit because they're based on technical considerations. It is useful to study lighting with open eyes. As mentioned in a previous chapter, when the CIE introduced an international trichromatic colorimetry and photometry system, known as the CIE System, in 1931 and illumination engineering societies were created, the more visual skills of the lighting designer declined. Design shifted from visually oriented people to technically oriented people. Architects and engineers now struggle to collaborate because of the the lack of training in architectural lighting design for architects.

Technical developments in lighting have been rapid, with more than 100 years of lighting research achieved through solid engineering expertise. General inflexibility in how research results are applied has been due partly to the fact that gas discharge lamps weren't amenable to adjustment at a reasonable cost. But the situation is changing rapidly, thanks to the endless possibilities of flexible LED light sources and electronic light control systems—not only in lighting design, but also in Darkness Design. The latest innovative area is smart urban solutions that integrate different urban elements into a complex smart grid, including lighting.

Modern technical tools for Darkness Design

It's relatively daring to write about modern LED lighting and electronic control systems, because by the time this book is published, its knowledge will already be partly out of date. The jungle of system innovations, both well-known and not so well-known, has become a clear problem even for experienced electrical engineers. Systems and components from different manufacturers and suppliers are often not mutually supportive or compatible, which has created uncertainty in design work. It's a curiosity in this field that lighting systems suppliers now hire electronics engineers instead of electrical engineers.

This book presents some brief examples of this rapid system development. The good thing is that everything you would like to create in a lighting project now can be realized. The suppliers' personnel are well trained, have excellent control over their own systems, and are happy to help.

Example 1 from 2010: New ULO–CT = 0 "Solar Brando LED"
Author and architect Oliver Walter wanted to modernize the previously presented Brando luminaire so that it included:

1) ULOR = 0 (in keeping with the ULOR–CT program)
2) LED panel
3) Solar panel
4) Flexible and freely selectable lighting control
5) Solar panel energy production routed back to the grid during the day

Figure 125: ULOR = 0

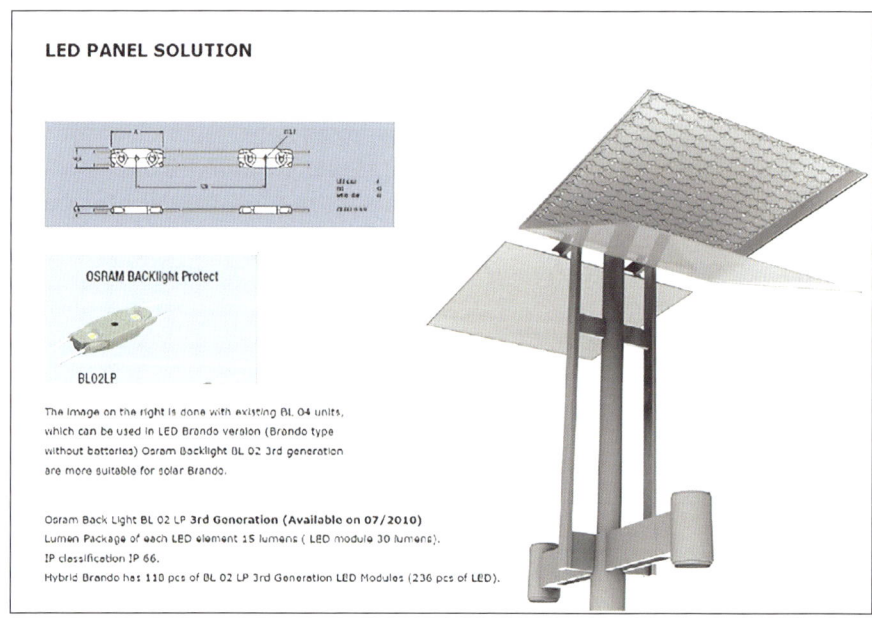

Figure 126: LED panel solution. Photo Julle oksanen and Oliver Walter

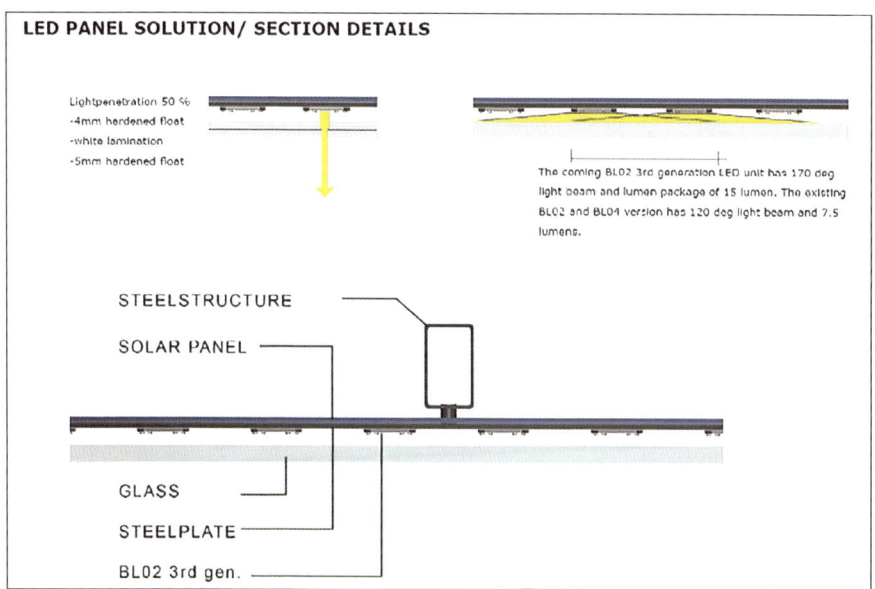

Figure 127: LED solution detals. Phot: Julle Oksanen & Oliver Walter

Automatic approaching control system for Brando

A study on electronic control of the Brando luminaire (designed in 2010 and already old) represents an example of first steps towards valuing biodiversity in the design of outdoor lighting. At the moment, the fastest developing sector in the lighting industry is lighting control management. The reasons for that include the possibility of huge energy savings, the ease of implementing such systems, and quick payback times.

The Solar LED Brando is equipped with an approaching-lighting dimming system. LED panels (9 pieces of Osram Prevaled Area-type plates per Brando) get their energy either from batteries inside the poles or the electrical network, depending on the battery load. An automatic switch takes care of choosing the right energy source. Direct current (DC) flows from the solar panels into the batteries. The batteries are connected to an inverter that turns DC to AC, which is suitable for use by the Brando luminaire.

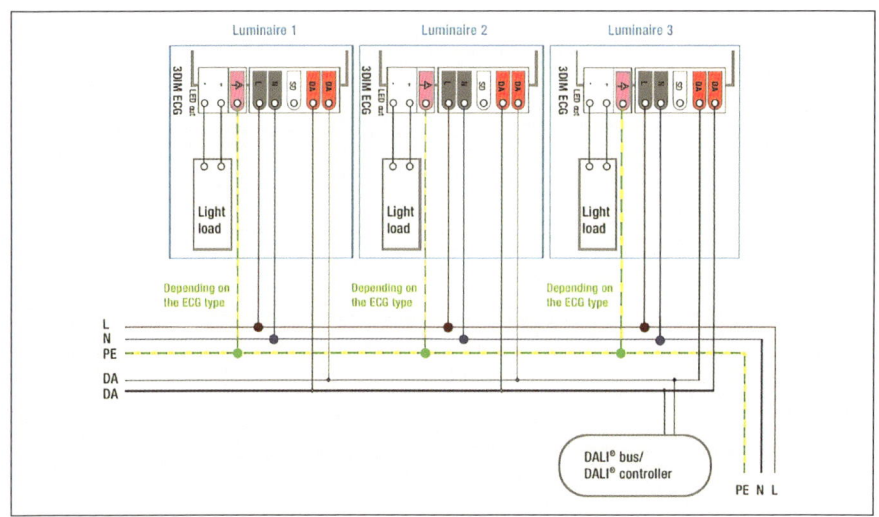

Figure 128: StepDIM activation for Brando. Photo: Osram

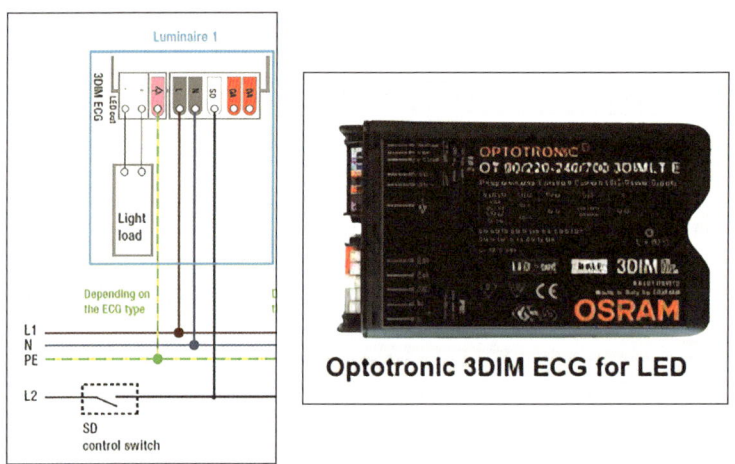

Figure 129 left and right: Photos: Osram

Electricity flows into the Optotronic control gear either from the street lighting network or Brando's own solar system inverter. When it detects an approaching pedestrian/cyclist or car, it sends an electrical impulse to the control switch SD. The Optotronic 3DIM ECG for LED component can be preset to the function desired. For example, when the path/road is empty, with no movement, the light is dimmed to 10% of maximum. When a person approaches, the light level slowly rises to 100% (which can be adjusted after testing the right speed). Some minutes go by (time adjustable) and, once there is no movement, the light drops again to the level of 10% of maximum.

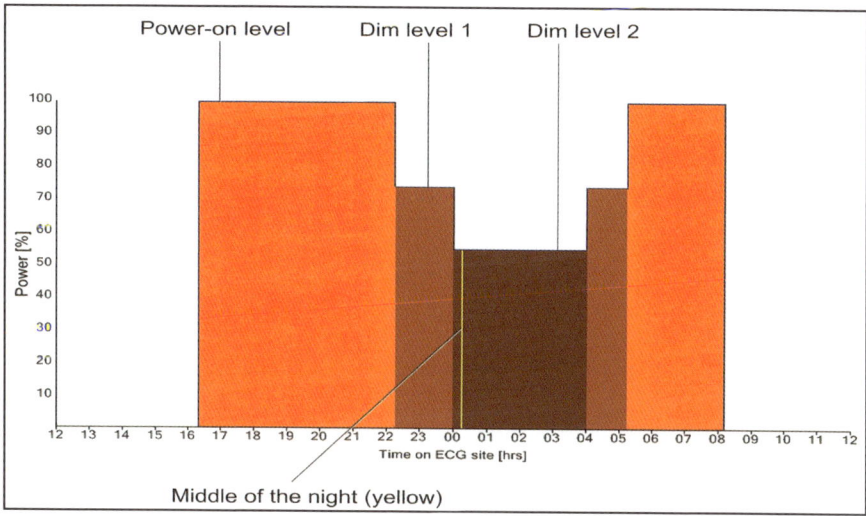

Figure 130: Photo: Factory setting for Optotronic 3DIM ECG for LED control gear. Between L1 and N mains is 230V. SD keeps power up (e.g. 100%) until its programmed fading model starts to fade lighting according to a programmed time, reducing it to the programmed dimming level (e.g. 50% of max). Fading times and power levels can be programmed freely. Photo: Osram

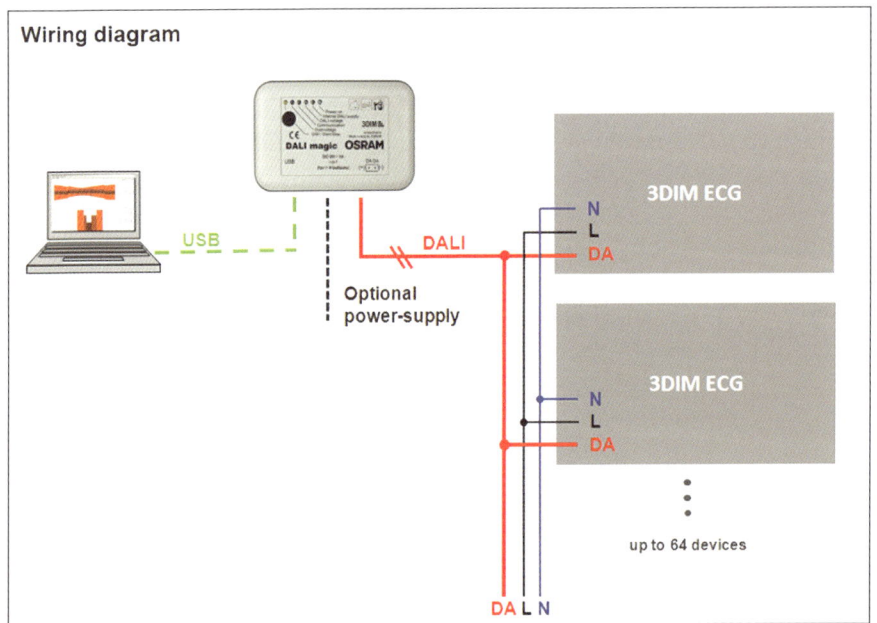

Figure 131: Optotronic can be programmed to work as desired by using the Osram Dali magic component and a PC. Photo: Osram

Figure 132: Engineer Olli Oksanen, from Osram, tests the Brando luminaire pre-settings for an LED driver with a computer. Computer + Osram Dali magic + 3DIM ECG + LEDs on the testing table. Photo: Julle Oksanen

Example 2: Technical solution for the "Light, Glass and Bird Behavior" curve, new façade lighting

A plethora of LED interior lighting control systems are available for buildings. However, implementing the "Light, Glass, and Bird Behavior" curve in urban lighting requires, among other control systems, an electronic device that automatically detects the changing times of dusk and darkness at particular longitudes and latitudes, on a daily basis.

The Astronomical Clock Switch with a weekly program is suitable for this purpose. For example, the Theben Selekta 171 top3 RC astronomical digital clock switch includes dozens of different functions. Devices with important functions for this purpose are:

1) Astro pulse, which can be used to control the building's interior lighting system to operate according to the control curve represented by a yellow dashed line. Channel 1, marked in red on the bird curve, controls the lighting level to 20% of the maximum (e.g. 40lx if the office 100% is 200lx).
2) Channel 2 gives a signal to the lighting control system to lower indoor lighting during Nautical Dusk, for example to 10% of the maximum (20lx).
3) Channel 3 gives a signal to the lighting control system to reduce indoor lighting during Astronomical Dusk to, for example, 5% of the maximum (10lx).

Each channel requires its own Selekta 171 device, because, despite numerous other device control functions, the device has only one astronomical time selection for three individual twilight zones.

The latitude and longitude of the building can be programmed in, using coordinates or choosing from a list of cities.

Once the device is preset, it will function automatically for as long as the device works. New presets can be made if necessary.

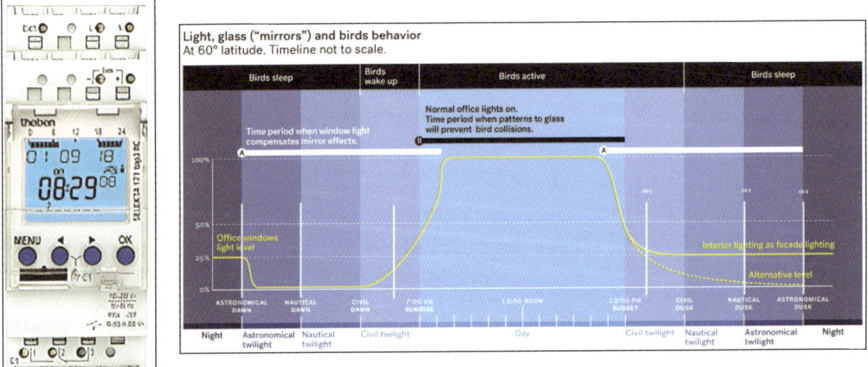

Figures 133 left and right: Left: Theben Selekta 171 top3 RC Astronomical clock switch. Right: Operating on-off channels marked with white lines.

Figure 134: Possible biodiversity-friendly Channel 1 lighting. Interior lighting serving as façade lighting is a poetic solution to the lighting dilemma. As Gaston Bachelard writes, "We are hypnotized by solitude, hypnotized by the gaze of the solitary house; and the tie that binds us to it is so strong that we begin to dream of nothing but a solitary house in the night." Richard von Schaukal rhapsodized: "Oh, light in the sleeping house." Photo: "Telenor building 'sleeping' in the Northern night": Jan Drablos

Example 3: Interact City/Signify

A century ago, engineers developed electric outdoor lighting without knowing anything about its devastating effects on biodiversity. Now engineers have developed an LED light source and suitable intelligent electronic systems that allow us to adjust this lighting to whatever we want. The original idea of all this was to save energy and integrate the technical elements of urban structures. Unknowingly, they created a technological solution to help protect biodiversity.

The exponential development of LED light sources and the versatile electronics that control them has sparked a rush to integrate urban lighting systems with other urban systems. Unlike before, these systems are not yet standardized along a common operating model, but everyone is striving to grab their share of the cake by developing their own concepts.

As cited in a 2018 report from the Population Division of the UN Department of Economic and Social Affairs (UN DESA), the world's urban population is expected to double by 2050. Addressing this global megatrend and the challenges inherent in that trajectory, the Signify Interact City portfolio of products and services has been designed to help create more equitable and sustainable cities while saving over 80% in lighting energy costs through deploying connected LED technologies. These savings can be invested in green funds, which, in turn, can support community development initiatives and quality of life improvements in deprived areas.

In a recent report from Guidehouse Insights (formerly Navigant Research), Signify was rated for the third year in a row as the global leader in both strategy and execution of smart street lighting. Guidehouse Insights defines "smart street lighting" as the use of advanced technologies to improve operational and management efficiencies with a connected smart city strategy.

As urban environments grow, change, and seek to better serve their residents, establishing the base of a smart city infrastructure through connected street lighting is a logical first step. Enhancement of owned or leased existing infrastructure is often simple, enabling rapid improvements in sustainability and neighborhood equality.

Interact City is Signify's connected lighting system for monitoring, controlling, and managing LED street lighting. Interact City enables cities to be "smarter," to reduce energy consumption and operating costs, and to become more livable and safer for residents. A set of application programming interfaces (APIs) makes it possible to integrate lighting with other management systems, as well as to develop services that that can make use of data collected by sensors embedded in the lighting system. Signify has successfully implemented over 2,000 projects in 58 countries. One example: Large-scale smart street lighting projects have been deployed for the New York Power Authority in the United States.

Global growth of intelligent lighting

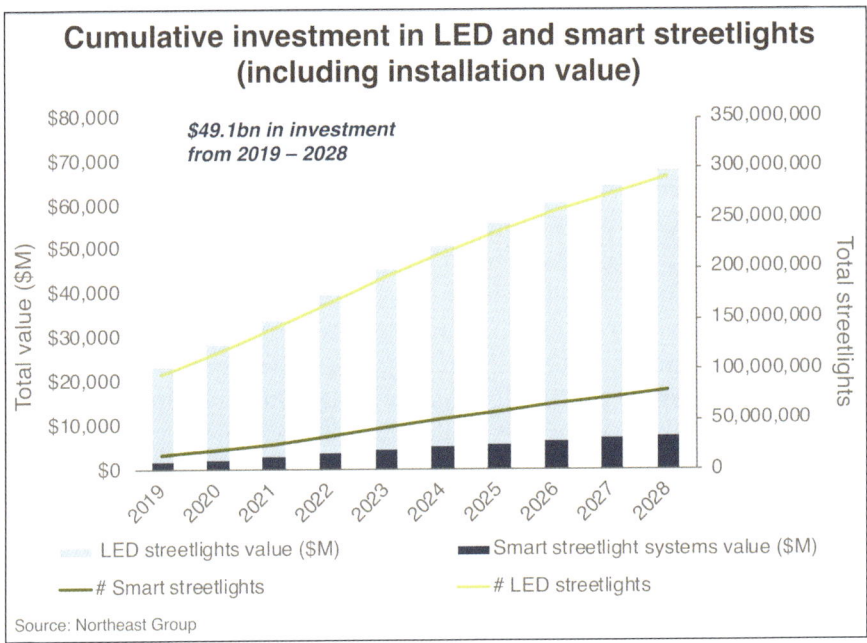

Figure 135: "Regional overviews / Global Smart Street Lighting and Smart Cities". Photo: Northeast Group, LLC.

Opportunity and Challenge

The lighting field as a whole is in the grip of intense, rapid change. Taking into account and integrating the research results of architectural lighting, technical lighting and biodiversity in the design of intelligent lighting systems is a demanding task. It is clear that massive and comprehensive training is needed in all professional areas if we want to have nocturnal urban spaces that value biodiversity and architecture and are equipped with high-quality technology.

We need to rescue ecosystems from light pollution and design tools for nocturnal darkness, to create pleasant and glare-free nocturnal solutions for everyone, animals included. Research results from all relevant parties (biology, engineering, architectural lighting, landscape architecture, vehicle lighting, road lighting, manufacturers, etc.) can finally be applied, with and without intelligent lighting systems.

BIODIVERSITY AND ARCHITECTURAL DARKNESS DESIGN TOOLS

The Darkness Design process

The role of concept in Darkness Design

Creating the right concept is a primary tool in a creative and successful process of Darkness Design. Without a clear understanding of the underlying concept, trying to beautify a city using darkness becomes a matter of luck.

The concept will arise out of a great unknown + personal semiotic connections + ideas + the collective creativity of the designer/design team. A concept can be created in minutes, or it may take a year, depending on how much creativity is sparked in the designer's brain or the collective brains of the design team, and how it can be drawn out to serve the task in real life. The darkness designer collects an idea library in his/her brain in the course of professional practice.

Concept brainstorming can be conducted alone or in a team. The designer or team members together put forward a variety of ideas; these ideas are evaluated and the best one selected to become the concept. When the concept is ready, it should be introduced to a representative of the client (a creative person who understands the importance of Darkness Design). Then the real design work starts, with the master plan design phase.

There are five different ways to approach conceptual design: 1. Thought-free; 2. Ideas from the past; 3. The analytic approach; 4. The gestalt approach; and 5. The intuitive unconscious/super-conscious approach, depending upon the designer's skills. The nature and special demands of the space to be designed also help to determine the conceptual focus.[24]

---

24 More on this topic can be found in the doctoral dissertation: https://aaltodoc.aalto.fi/handle/123456789/27886

Darkness Design master plan

The master plan occupies a strategic place in the Darkness Design process. Designers need to find darkness hardware and software solutions to implement their concept on a practical level. If the right solutions cannot be found, it may be necessary to change the concept. In some cases, it may even be necessary to create a new concept.

Accumulating working experience in the real world and in real projects develops a darkness designer's perception of what can and cannot be implemented. Sometimes even a great concept ends up staying in the realm of dreams, but often a small change can save the day and produce a high-quality result.

The master plan phase consists of selecting lighting fixtures, lamps, testing out construction solutions, making darkness calculations, creating computer images, etc. This phase excludes detailed design elements such as working drawings, construction detailing and luminaire integration into structural elements. It's a good time to clarify what is in the offer, to avoid unclear situations later about what is included in the work and what is not.[25]

Darkness Design details

Big projects need a professional coordinator, especially when construction is moving ahead quickly. For me, a great example was the Telenor project. In that case, the project coordinator was a talented and experienced project leader, Jan Drablos of Multiconsult Norway. He transferred our Darkness Concept and Master Plan information to the electrical engineering consultants, whose task was to add lighting solutions as part of the electricity drawings prepared for the electrical contractor. In the detailed design phase, the Darkness Designer may have only a small role in a huge project. In the implementation phase, Jan Drablos from Multiconsult decided to take full control of the whole Darkness Design and procurement, even though it was a technical turnkey contract. This was to avoid being cheated on types of fixtures (which is a common problem all over the world).

Multiconsult decided on all the manufacturers and types of fixtures. Testing was often carried out before making a decision on type and installation. The electrical engineering consultant team was instructed to perform

---

[25] More about the topic can be found in the doctoral dissertation: https://aaltodoc.aalto.fi/handle/123456789/27886

the final detail design according to the darkness designer's concept, using Multiconsult's choice of luminaires, with great success!

Our darkness project was completed successfully with only one luminaire type, designed by ourselves (Vesa Honkonen & Julle Oksanen). Notor was manufactured by the international luminaire manufacturer Fagerhult Belysning AB. Fagerhult is one of Europe's leading lighting companies, with 2,200 employees in 20 countries. Fagerhult develops, manufactures and markets innovative and energy-efficient lighting solutions for professional indoor, retail and outdoor environments. Telenor has over 8000 meters of Notor luminaires. (More about this topic in Appendix 1)

The design process according to Richard Kelly

Perception psychology

"Perception psychology" refers to the psychology of a mental process in which sensory and emotional data is organized logically or meaningfully in the observer's mind. The process provides the basis for understanding, learning, and knowing, or for motivating a particular action or reaction.

Richard Kelly's lighting ideas apply perception psychology to architectural outdoor lighting design. Kelly (1919-1977) was a pioneer of qualitative design who incorporated ideas from perception psychology and theatrical lighting into a uniform concept for lighting design. He broke away from the rigid constraints of using uniform illuminance as the central criterion for lighting design in the "All surfaces white" era of architecture. He replaced the question of lighting *quantity* with the question of individual *qualities* of light. These qualities correspond to a series of lighting functions, in turn geared towards the perceptions of the observer. Three of these functions are critical to successful Darkness Design: "ambient luminescence," "focal glow," and "play of brilliants."

Figure 136: Richard Kelly. (Richard Kelly's archives).

Ambient Luminescence

Kelly called the first and fundamental form of light "ambient luminescence." This is the element of light that provides general illumination of the surroundings; it ensures that the surrounding space, its objects and the people there are visible. This form of lighting facilitates general orientation and activity. Its universal and uniform orientation means that it largely follows the same lines as quantitative lighting design, except that ambient luminescence is not the final objective but just the foundation for a more comprehensive lighting design. The aim is not to produce blanket illumination, or "one size fits all" lighting at the supposed optimal illuminance level, but to have differentiated lighting that builds on the base layer of the ambient light.

Figure 137: Ambient Luminescence, theoretical display. Photo: ERCO

Figure 138: Ambient Luminescence, case study. Photo: Julle Oksanen

Focal Glow

Kelly termed his second form of light "focal glow." This is where light is given the express task of actively conveying information. The fact that brightly lit areas automatically draw our attention comes into consideration with focal glow. By using a suitable distribution of brightness, it is possible to order the wealth of information contained in an environment. Areas containing essential information can be emphasized by accented lighting, whereas secondary or distracting information can be toned down by applying a lower lighting level. This facilitates a fast and accurate flow of information, whereby the visual environment is easily recognized in terms of its structures and the significance of the objects it contains. This applies equally to orientation within the space (e.g., the ability to distinguish quickly between a main entrance and a side door) as it does to emphasizing certain objects, such as when presenting goods for sale or when highlighting the most valuable sculpture in a museum collection.

Figure 139: Focal Glow, theoretical display. Photo: ERCO

Figure 140: Focal Glow, case study Howard M. Brandston Lighting The Lady. Photo: Brandston Partnership inc.

To arrive at light differentiation, Richard Kelly came up with an arrangement of light he called "focal glow," in which light actively conveys information. Brightly lit areas automatically draw our attention, so areas with essential information (such as main entrances, or important objects) can be emphasized with accented lighting while secondary or distracting information can be toned down with a lower lighting level.

Play of Brilliants

The third form of light, "play of brilliants," results from the insight that light not only draws our attention to information but can also represent information in and of itself. This applies to the specular effects that point-of-light sources can produce on reflective or refractive materials. The light source itself can be brilliant. This "play of brilliants" can add life and ambience, especially to prestigious venues. The effects traditionally produced by chandeliers and candlelight can now be achieved in a modern lighting design by the targeted use of light sculptures or by creating brilliant effects on illuminated materials.

Figure 141: Play of Brilliants, theoretical display. Photo: ERCO

Figure 142: Play of Brilliants, case study; Gateshead Millennium Bridge. Lighting design Speirs + Major. Photo: Andrew Curtis, licensed under CC BY SA 2.0

Real-life example

A good lighting design project to demonstrate Kelly's three principles is the Kilden concert and theatre hall in Norway. Designed by ALA-architects, it has a lighting concept by Julle Oksanen Lighting Design Ltd.

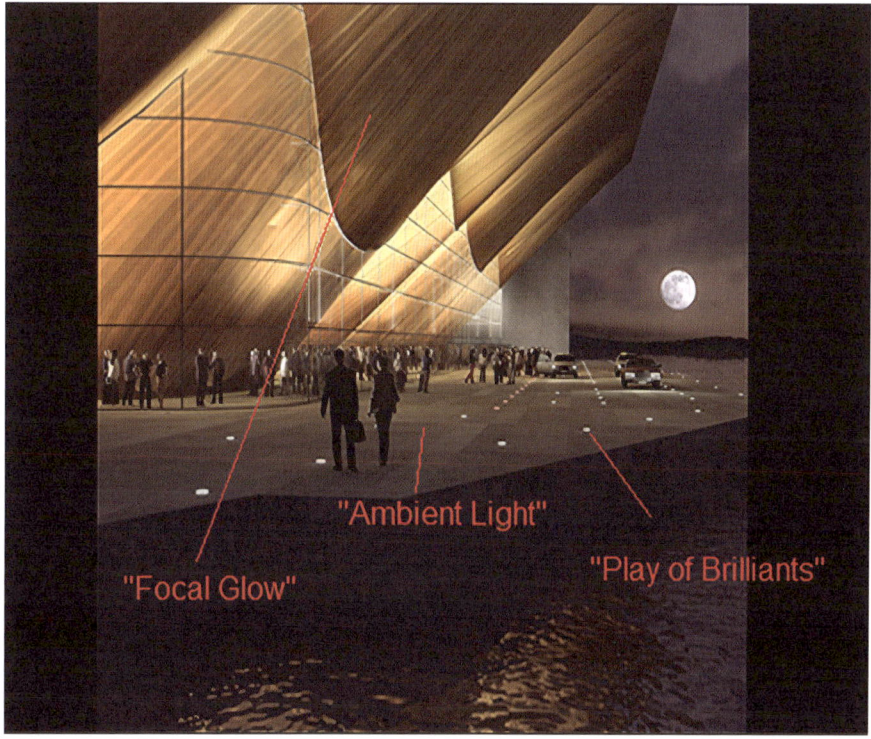

Figure 143: "KILDEN". Architect and CAD manipulation. Photo ALA Architects Ltd

"Ambient Light" : The glowing facade produces horizontal light on the ground. The horizontal light meets the minimum engineering and code standards for levels of light necessary for factors such as safe movement. Minimum values are acceptable due to other fixed lighting in the area, and 'live lights,' coming from cars for example, will provide more light. In other words, if maximum lighting were provided, additional 'live lights' would raise lighting levels far beyond the amount necessary, creating light pollution.

One of the most important goals in architectural Darkness Design is to avoid glare—light's worst enemy. In this case, heuristics helps by applying one's own experience of measuring various luminance values. As a rule of a thumb, contrasts are strong, and there is no veil over the human retina if the maximum brightness of the light distribution surface from all perspectives is at the level of lunar brightness.

"**Focal Glow**" : In-ground luminaires differentiate the facade of the building by creating a glowing wooden surface. Because of its attractive, statuesque nature, the curved wooden facade was selected as a "focal glow" element of the landscape lighting, creating a glowing landmark. By switching various buried luminaires on and off, it can be used as an information wall. One example of this informative quality is that more light is focused on the entrance of the theatre when there is a performance than when there is not.

"**Play of Brilliance**" : In-ground luminaires (stars on the ground). This landscape lighting solution was based on buried luminaires creating "stars on the ground." This exciting solution removes the need for ugly and disturbing technical lighting poles and glaring floodlights.

Figure 144: Punkt Festival 2012: The Venue

Hopkinson's Scale of Apparent Brightness

One of the five pillars of Darkness Design

As a young engineer, R.G. Hopkinson was assigned his first task, a street lighting project. When measuring road sign luminance values, he wondered why the meter did not show what he saw with his own eyes. The nighttime measured luminance was 34 cd/m^2, although the sign seemed every bit as bright as it looked in daytime, when the "brightness" of the sky was more than 100 times greater. (The luminance of a street sign during bright daytime is easily 3500 cd/m^2.)

Figures 145: The traffic sign on the right is 100 times brighter than the one on the left. Photo: Julle Oksanen

Hopkinson created Scales of Apparent Brightness based on this observation, which has pushed Western-style shadow treatments in the right direction. J.M. Waldram used his "Designed Appearance" technique in the lighting of many English cathedrals—a practical implementation of Hopkinson's Scales of Apparent Brightness.

Unfortunately, engineers working on a large scale did not know how to follow up after Hopkinson completed his research and made the results ready for use. This book aims to properly honor Hopkinson's research results.

"Lighting is a technology in which the relation between physical energy and the visual sensory reaction is of prime importance. Photometry would be impossible without the basic Relative Luminous Efficiency function for the standard observer. This basic stimulus-sensation relation enables us to build a scale of photometric units in terms of the energy of stimulus," Hopkinson wrote in 1957.

A very precise scale, he wrote, "is hardly feasible" because:
- Our visual mechanism at any one moment can register only a limited range of energy stimulus. A small amount of energy reaching the eye may cause no sensation of light (it is below the threshold of sensation).
- Too much energy saturates the vision mechanism.
- The increment of energy, which causes a given apparent change in brightness, varies systematically in a non-linear manner (equal changes in luminance do not produce equal changes in subjective brightness).
- There is uncertainty and hysteresis in stimulus-sensation relation.
- Human beings cannot assign an exact numerical value for visual sensations.

However, Hopkinson outlined methods that could be used to derive an Apparent Brightness Scale:
- Contrast Scaling
- Luminosity Photometer
- Method of Direct Estimation

In the conclusion of his scientific article, Hopkinson states that his paper is only the beginning, not the end, of the story. It shows, I think, that a useful engineering scale of apparent brightness can be derived by various methods, including direct estimation. The effects of adaptation can be allowed for.

Hopkinson writes that much more work is necessary before a final scale can be proposed, but in my personal experience of using his Scale of Apparent Brightness in my own projects, the existing basic research is sufficient. I have two reasons for this opinion: 1) Ambient lighting levels are always low in lighting design projects, where contrast ladders work more exactly than at high ambient lighting levels; and 2) According to S.S. Stevens' research, human beings cannot assign an exact numerical figure for observed lighting levels.

Hopkinson nevertheless hoped that CIE could turn its attention to solving the problem. Unfortunately, lighting research activities in CIE have dispersed on a large scale to address a great many new challenges in the lighting world, and Hopkinson's Scale of Apparent Brightness has been left behind as a valuable tool for possible future use.

This book explains the use in practical applications of modified contrast ladders based on Hopkinson's Scales of Apparent Brightness, with the help of the Pragmatic Theory of Truth. Hopkinson was a scientist and used scientific procedures. His study of the Scale of Apparent Brightness is based on scientific observations and scientific nomenclature (such as "luminance stimulus" and "apparent brightness on an arbitrary scale").

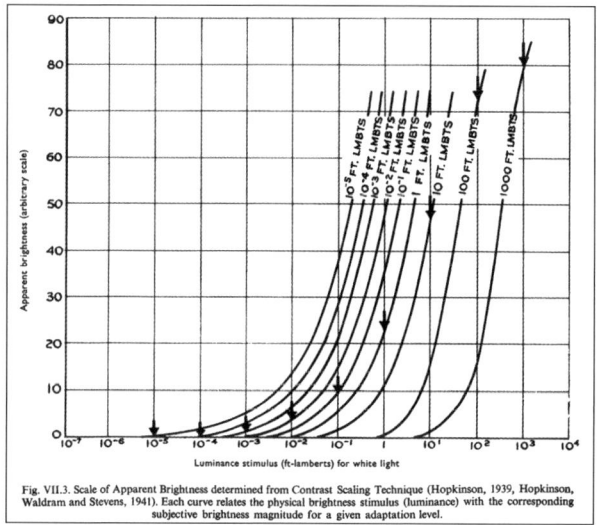

Figure 146: Scale of Apparent Brightness. The diagram is scientific and too complicated to be used for practical Darkness Design projects. To be able to use these scientific results in real-life projects, we need to employ a key: the Pragmatic Theory of Truth, an instrument that straddles the line between scientific theory and practice.

Heuristics and pragmatism as instruments

One problem is "shadowing" that may even prevent the use of Hopkinson's Scale of Apparent Brightness diagram in practical architectural Darkness Design. Research tests were implemented in "black and white,"[26] but in real life we have colours (such as building facades, trees, monuments, etc.). Another problem was the difficult scientific nomenclature, which was complicated for practical lighting designers (engineers, architects, etc.) to understand and use. The Pragmatic Theory of Truth can help in implementing in architectural Darkness Design this

---

26 Hopkinson 1957c

valuable diagram, with all its scientific lingo and practical uncertainties when defining perceived brightness values and linearly measured luminance values.

The Pragmatic Theory of Truth can be used to develop a pragmatic, collectively accepted tolerance between scientific research results and practical values in real-life projects, thus saving huge amounts of energy in public lighting. Hopkinson's Scale of Apparent Brightness can be modified to serve the purpose of practical Darkness Design, especially in the case of low lighting levels, where contrast ladders work perfectly without saturation of sight.

Practical use of Hopkinson's Scale

First, let's alter or replace scientific terminology with practical terminology, using the Pragmatic Theory of Truth:

1) When we combine Hopkinson's scientific Apparent Brightness Scale with the pragmatic truth factor, we arrive at "relative brightness," which in practical terms means "visible units." The change between each vertical unit on the scale (1-2-3-4-5-6-7 … 90) is barely perceptible.
2) When we combine Hopkinson's scientific "luminance stimulus" (tested with a square field patch subtending 3 degrees to the side at the subject's eyes) with the pragmatic truth factor we get the "luminance of the object" (façade, tree, monument etc.).
3) For practical reasons, we have to change the old-fashioned Footlamberts (fL) units from the original Hopkinson's curve to cd/m^2. $1 fL = 3.43\ cd/m^2$
4) A green, 25 cd/m^2 line is also marked and classified as a limiting average maximum façade luminance value (luminance of object) in the environment classification E4 (Cities and Towns, with strong environment lighting).

Defining the adaptation level as ambient luminance in real-life architectural Darkness Design projects can, in some cases, cause problems, especially if the ambient light in the observer's view is not solid. For example, in a dark forest, ambient light is practically solid and easy to measure with a luminance meter, but in urban areas, for example, there are areas of differing brightness. In cases where the ambient light is not solid and easy to define and measure, the designer can take many measurements and use an average value of these. Today, in extremely busy city centers, the average ambient brightness may rise so high that it is no longer practical to use this method at all. But the method can be used to present new values that guide the design of darkness.

Pragmatic example 1: Flashlight test

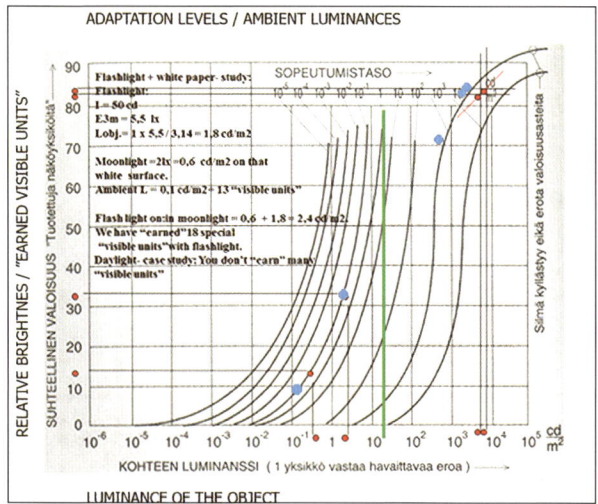

Figure 147: A pragmatic flashlight test with modified contrast ladders, according to Hopkinson's Scales of Apparent Brightness. The red dots represent a quickly done heuristic calculation by Julle Oksanen on a real-life project, and the blue circles are the results of Dr. Kevin Mansfield's scientific calculations. Photo: Julle Oksanen

The forest-flashlight-white paper test serves to clarify the use of modified contrast values and also opens the door to Hopkinson's scientific world. The test is very simple and educational: Two people go to the forest; one has a flashlight and the other a sheet of white A4 size paper.

Assume the white sheet of paper is lit by a flashlight against the background of the forest at night and during the day.

Night
Assume reflectance (ρ) of background forest = 0.2; that of the paper to be 1.0 (although a better estimate would be 0.8).

Assume illuminance due to the moon = 2lx

Assume flashlight of intensity = 50cd at a distance of 3m from the paper.

$E = I/d^2$ and therefore the vertical illuminance, Ev on paper due to flashlight = 50/9 ~ 5.6lx

From $L = \rho E/\varpi$, luminance of paper, Lp = 1 x 5.6/ϖ = 1.8cd/m² and luminance of background, Lb = 0.2 x 2/ϖ = 0.1cd/m²

Day
Assume illuminance due to daylight = 10 000lx (strictly speaking on the horizontal plane, but a reasonable estimate for the vertical plane).

From L = ρE/ϖ, luminance of paper (*without* flashlight), Lp = 1 x 10000 /ϖ = 3183cd/m², luminance of paper (*with* flashlight), Lpf = 1 x 10006 /ϖ = 3185cd/m2 and luminance of background, Lb = 0.2 x 10000/ϖ = 637cd/m²

Adaptation luminance
There is constant debate about what value should be taken for adaptation luminance. Waldram, for example, suggested taking the average luminance in the field-of-view within 10° of the area of vision. In this example (ignoring the luminance of the paper), a reasonable approximation can be an adaptation luminance of 0.1cd/m² at night and 637cd/m² during the day.

The object, background and adaptation luminances can now be plotted on the metric version of Hopkinson's Apparent Brightness Scales prepared by the author (approximate **blue** circles).

So at night the Apparent Brightness of the paper against the forest background is 33-8, a brightness ratio of about **4:1**.

During the day, the Apparent Brightness of the paper with and without flashlight against the forest background is 85/84-72, both around **1.2:1**.

"This explains why the effect of the flashlight is so marked compared to the situation during the day."[27]

> **Pragmatic example 2: Practical use of Hopkinson's Ladders, Cathedral in Arras, France**

In 2000 the architect Vesa Honkonen was invited to participate in a lighting design competition to create an exterior lighting design for the medieval town of Arras, France. The lighting design was completed together with the author. Contrast ladders, a term that reflects the gradual increases in "ambient luminances" in Hopkison's Scale of Apparent Brightness, were used to define the target luminance value for various parts of the project, e.g. for the Arras cathedral (marked number 12 on figure below).

The main lighting design philosophy was, on the one hand, to allow darkness in the central parts of each plaza (marked in gray) and to illuminate the peripheral regions (yellow). On the other hand, the idea was also to dim the lighting for a short distance at the approach

---

27 Mansfield 2016

to the plaza. The observer would arrive at the plaza from a tiny, dim street and thus enjoy entering the plaza and seeing its illuminated facades. This would have been easy to do because of the modest classifications of the tiny streets of Arras.

Figure 148: Arras town plan. The cathedral, numbered 12 on the image, was also introduced in the competition entry. Photo: Vesa Honkonen Architects, architect Mari Koskinen.

Figure 149: Computer simulation of the cathedral façade lighting, according to lighting designers Vesa Honkonen & Julle Oksanen. Photo: Architect Oliver Walter

Figure 150: Lighting design elements with different luminance values were marked on a black & white photo as A, B, C, and CN. Photo: Oliver Walter

12. CATHEDRALE							
	LUMINANCE DE LA SURFACE	FACTEUR DE LA REFLEXION DE LA SURFACE	NIVEAU D'ILLUMINATION	TEMPERATURE DE COULEUR		OBJET / ARRIERE-PLAN	LUMINOSITÉ RELATIVE
A	6 cd/m2	0,3	60 lx	3000 K	1	A / B	27
B	4 cd/m2	0,3	40 lx	3000 K	2	C / CN	30
C	2 cd/m2	0,3	20 lx	3000 K	3	A / CN	45
CN	0,1 cd/m2	-	-	-			

LUMINOSITÉ RELATIVE 15=contraste doux 30=contraste claire 45=contraste forte 60=contraste tres fort 75=contraste gênant (± 5 unités)
CN=ciel de nuit

Figure 151: Different luminance values were determined by estimating from a computer simulation. Illumination values for various elements were defined by the reflection factor of the surface of the cathedral (the reflection value being estimated as 30%). Because of the "whispering lights" type of luminance values desired, a warm color temperature was selected.

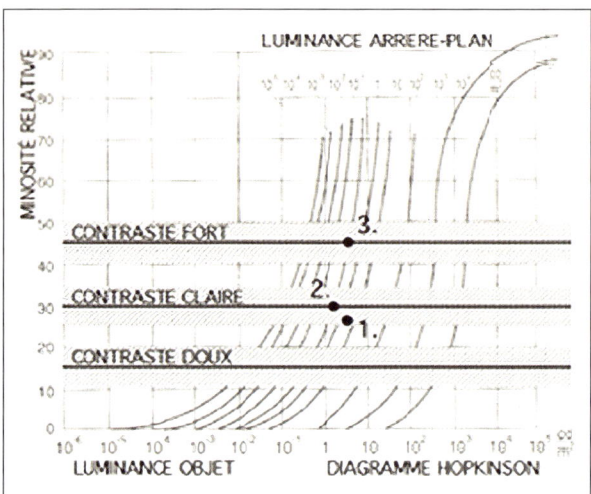

Figure 152: Contrast ladders were used to determine contrast values, which of course also influenced the luminance values of different lighting elements. The highest "visible units" values were between the dark sky, marked as CN (ambient luminosity, Lambient = 0.1 cd/m^2) and illuminated pylons, marked as A (Lpylon = 6 cd/m^2). Between these elements "earned visible units" amounted to 45.

We used Hopkinson's Scale of Apparent Brightness as part of the lighting composition. The "whispering lighting" on the façade is peaceful and does justice to the nighttime architecture. Unfortunately, contrast ladders are not commonly employed because of the lack of lighting education.

HEURISTIC APPROACH

It's good to remember that the strict, methodical practices required of professionals don't necessarily sustain more general social aspirations. Developing new methodologies helps consolidate knowledge within a specific field, but it creates a barrier between experts and amateurs, and thus makes the field of study immune to external feedback and pressures. Although it's natural for settled patterns to develop in various specializations such as technical fields, health sciences, "design research," etc.[28], this isn't helpful to a field that needs to involve diverse actors.

---

28 Cross N. 2007 *Designerly Ways of Knowing*. Basel: Birkhäuser

Creative, light-pollution free architectural Darkness Design and research requires input from biologists, architects, interior designers, industrial designers, physicists, anthropologists, engineers, and so on. This demands a transformation in design that moves the field from methodology towards heuristics.

Turning attention to heuristics and finding that it's possible to achieve a good-enough result in architectural Darkness Design doesn't mean abandoning the systematic pursuit of knowledge, but only the pursuit of invoking universal technical lighting arguments. "In this sense, the heuristic approach is close to pragmatism."[29]

Pragmatism sees knowledge as a part of practice and emphasizes adopting methods derived from trial and error. Truth and knowledge are compared with that which "works." But as the social theorist Stephan Fuchs points out, it's not about how a single part functions, but the whole system.[30]

Although heuristics is not praised in scientific debate, it has been shown to be a pivotal part of problem-solving for both experts and laymen.[31] It is helpful in generating and transmitting ideas, and in coping with problematic situations. The heuristic method is rarely deployed in the technical lighting design process, where decisions are based on standards, norms, and recommendations. But in the architectural Darkness Design process, the heuristic approach is a very important tool, after choosing a conceptual approach and creating a new darkness metaphor.

Approaching in darkness

Gradations of darkness are the philosophical gateway to achieving intriguing, aesthetically successful darkness solutions. But new architectural Darkness Design metaphors and different conceptual processes must keep the aspects of safety and comfort in mind.

Numerical lighting values needed to ensure pedestrian safety have been established with the help of scientific research.

Mandatory facial recognition distance and minimum semi-cylindrical illuminance values for lighting in outdoor spaces were studied by anthropologist E.T. Hall, architect Jan Caminada and engineer Wout van Bommel in

---

29 *The Quest for Certainty: A Study of the Relation of Knowledge and Action.* New York: Minton, Balch & Company.
30 Fuchs 2001
31 Dreyfus, H.L. & Dreyfus, S.E. 2005 "Expertise in real world contexts". Organization Studies 26: 779-792. Kahneman,D. & Tversky,A. & Slovic, P. *Judgment under Uncertainty: Heuristics and Biases.* Cambridge,UK: Cambridge University Press, 1982.

the 1980s. Their research shows that measured semi-cylindrical illuminance values required were so low that tolerances and variations between different test observers make it quite possible to apply Darkness Design, instead of the existing practice of technical lighting design. Successful darkness research results ranged between Esc = 0.8 and Esc = 1lx.

"With our senses, we are able to detect certain stimuli in our surroundings. But most go unnoticed, for example large parts of the electromagnetic spectrum. So, the 'picture' our brains make of our surroundings remains limited—and the world we experience with our senses is not identical with the physical world," the researchers suggest. "If we had different or additional senses, the world would look very different. Visual perception is a process driven by sensation, with its outcome dependent on judgments based on the perceiver's situational experiences. Seeing is an intellectual exercise strongly influenced by perceptions and cultural experiences. It may be expressed in several ways, among them, verbally and pictorially."[32]

Studying the physiology of perception is important, because it allows us to understand how color and levels of light lead us to perceive our environments as legible and safe. Hall, Caminada and van Bommel studied how people behave in the dark and found two important elements relevant to lighting designers:

The distance required for recognition between people approaching each other in the dark is 3.5 meters (approximately 12 feet). In public spaces, there should be enough light to enable people to recognize each other at this recommended distance.

Personal zones:
- **Intimate zone: 0 – 0.5m**: Typical hugging and kissing distance.
- **Personal zone: 0.5 – 1.2m:** Typical close discussion distance.
- **Social consultative zone: 1.2 – 3m:** Typical communication distance.
- **Mandatory recognition distance begins at 3m:** Meeting comfortably in darkness.
- **Evasive/defensive action possible = 3.5m.**
- **Close zone: 4 – 10m.** Relaxed movement.

---

32 Science Center Spectrum, 2008

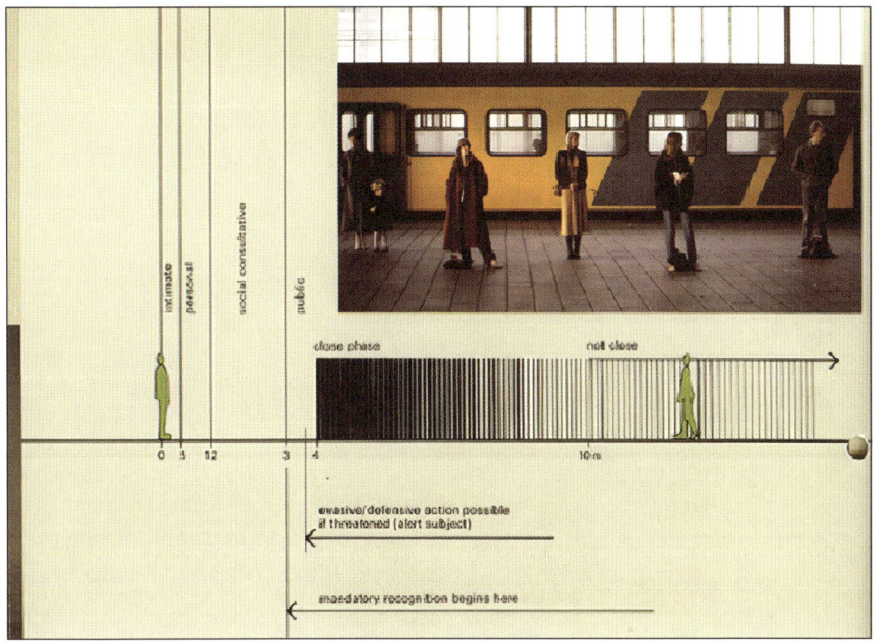

Figure 153: Personal zoning, according to research. Photo: (Philips ILR 1980 /3)

Recognition distance:

To meet this recommendation, the semi-cylindrical Illuminance value Esc should be higher than 0.8 lux, or approximately 0.08 foot candles.

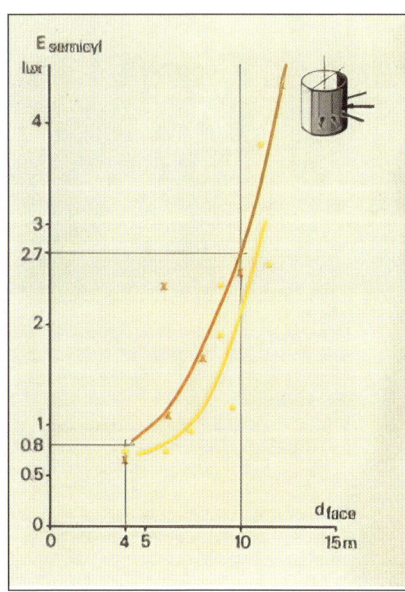

Figure 154: Esc- value required for face recognition when approaching in darkness. Photo: (Philips ILR 1980 /3)

Semi-cylindrical illumination values had coinciding and coherent results. Other values had some contradictions. Semi-cylindrical illumination value is understandably important because faces generally follow this basic shape. Variations in heads and faces didn't influence the result. Lighting design software programs automatically calculate this value using an average face height of 1.5m Escmin = 0.8 lx (at selected places Esc = 3 lx).

Sample computer calculations

Futuristic Dream City 2030

Figure 155: Calculation example: City 2030. Design Julle Oksanen. Photo: Oliver Walter and Daniel Silberman

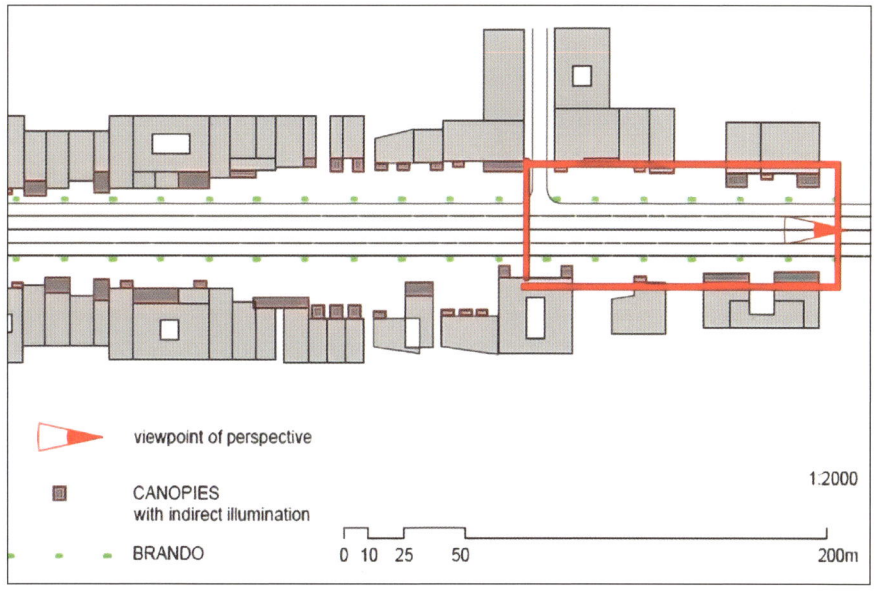

Figure 156: Calculation area. Photo: Oliver Walter

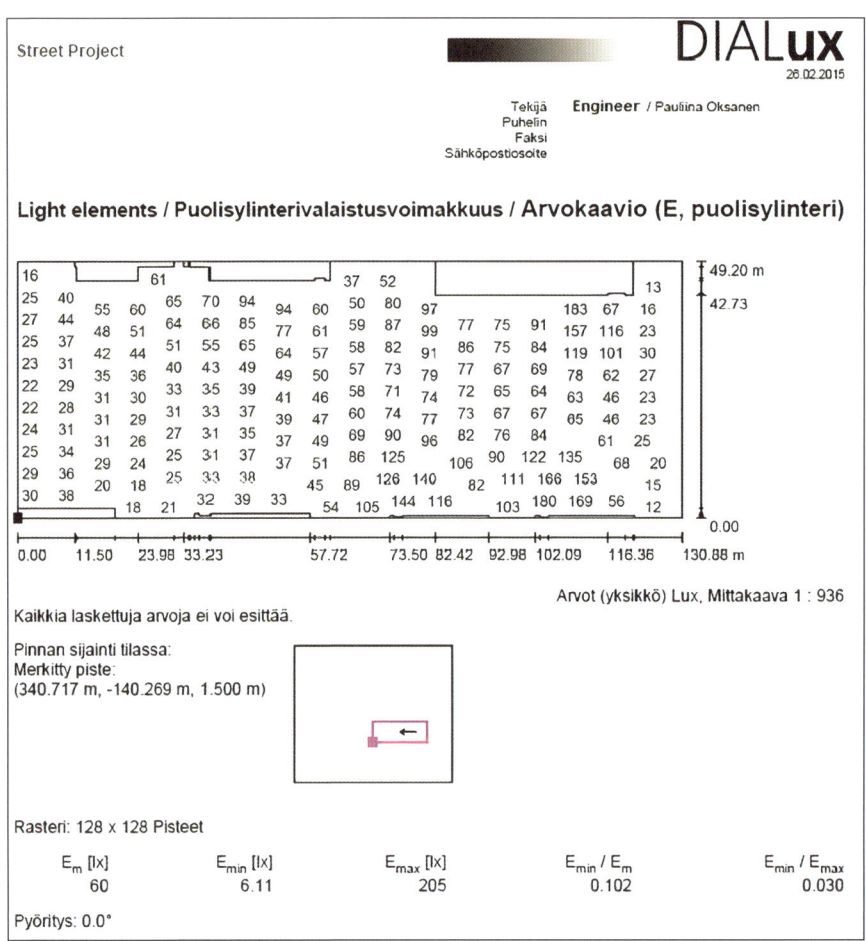

Figure 157: Average Esc value is very high. Esc = 60lx allows for dimming LEDs down to 90%. Then Esc = 6lx and facial recognition distance is approximately 15m. The possibility of producing graduated darkness is excellent, and the result is interesting and inviting. Photo: engineer Pauliina Oksanen

Approaching test:

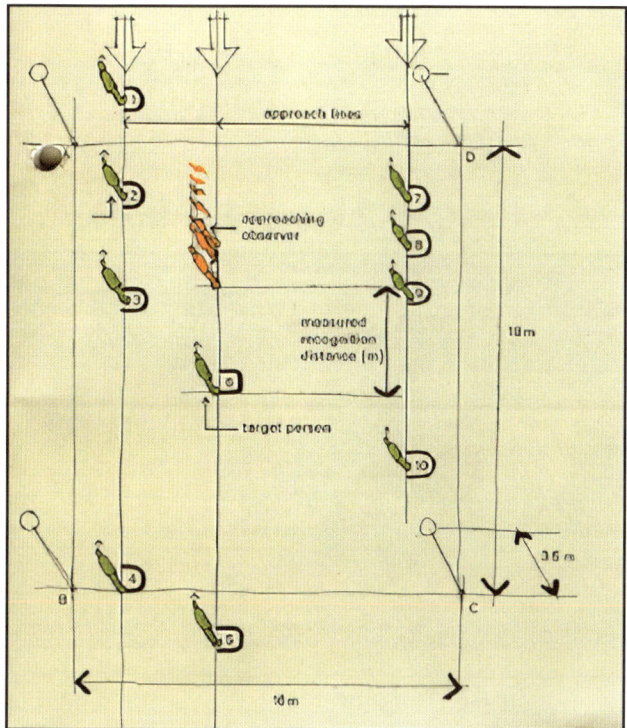

Figure 158: Approaching test implemented in Eindhoven. Photo: (Philips ILR 1980 /3)

This study was made in Eindhoven, Holland, where Philips is headquartered. Both Wout van Bommel and Jan Caminada worked for the Philips lighting division in the 1980s. Personnel from one of the Philips lighting departments were taken to an outdoor demonstration laboratory. The lighted area of the post-top fixture's distribution surfaces and the volume of light (influenced by candle power and brightness of the luminaire surface) could be altered in order to create a variety of lighting conditions. Each person was tested as other persons approached. The lighting values were adjusted each time and the test person was required to shout "stop" when he or she recognized the face and attitude of the approaching person. The distances were measured against the lighting values and also Ehorizontal, Evertical, Esemicylindrical, and Ecylindrical. Based on the results of this study, recommendations were made to produce uniform lighting values for public areas.

Fixtures used and their glare rating:

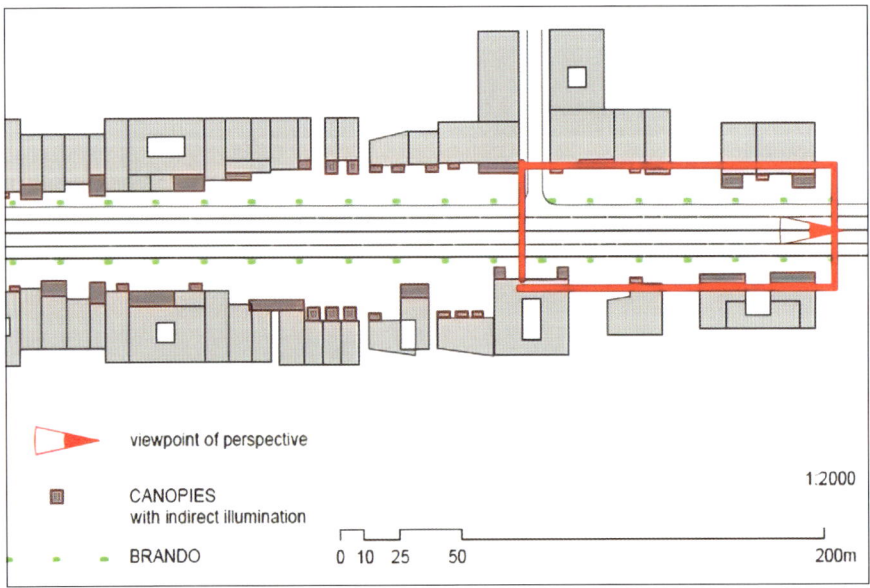

Figure 159: Principal drawing for LA 0.25- value. Photo: (Philips ILR 1980 /3)

If the fixture's surface is brighter than the recommendation allows, it affects recognition distance, because glare veils the retina. If it's too bright, the test subject loses the ability to distinguish contrasts. This actually shortens the recognition distance. If the distance is shorter than 10 feet, at least one person will typically change their walking direction because this situation causes uncertainty, and the space may be perceived as unsafe. Creating appropriate recognition zones is valuable for relaxed and enjoyable movement in darkness. The fixture glare value calculation formula is:

LA 0.25
where,
L = measured luminance between angles of 85-90 from vertical (bright area section of the luminaire light distribution surface in cd/m²),
A = bright fixture light distribution section in m².
The maximum values depend on the height of the fixture:

$LA^{0.25}$ =   1250, h < 4.5m
$LA^{0.25}$ =   1500, 4.5m < h < 6m
$LA^{0.25}$ =   2000, h > 6m.

Dream City 2030 glare values allow for implementing the perfect Darkness Design option and "whispering lights."

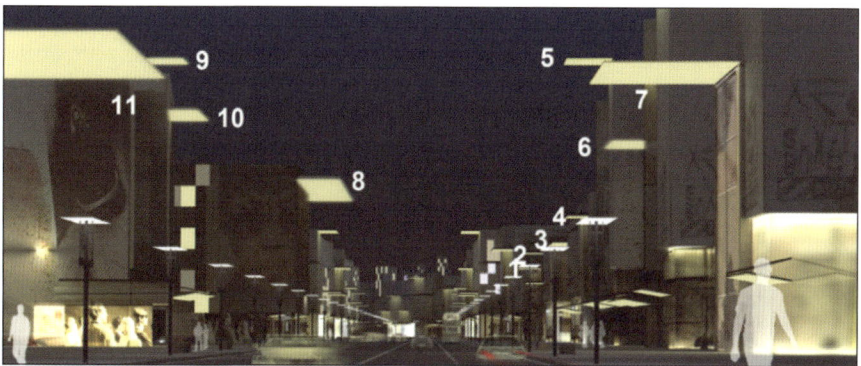

Figure 160: Fixture numbering for LA ^{0,25} calculations in Dream City 2030 case with Brando fixtures. Design Julle Oksanen: Photo: Oliver Walter

Fixture	Ara A (m²)	Hight (m)	Luminance (cd/m²)	LA^{0,25}
1	10	12	100	97
2	40	15	100	137
3	10	12	100	97
4	40	12	100	137
5	60	15	100	151
6	10	12	100	97
7	60	10	100	151
8	20	15	100	115
9	10	12	100	97
10	80	10	100	**162**
11	80	10	100	162

Sample calculus: Fixture number 10:
Area seen from angle 5 degrees from vertical:
A projection = A x sin 5
A projection = 80m² x 0,087 = 6,96 m²
Luminance of emitting surface = 100 cd/m²
LA 0,25 = 100 x 6,96 0,25 = 162,4 <<< 2000 (which is already a great value)

NOCTURNAL DARKNESS DESIGN, CREATIVITY STRUCTURES AND BIODIVERSITY

About shadows and nocturnal darkness

Professional Darkness Design should be considered the cornerstone of nocturnal design because:
1) **It protects biodiversity.**
2) It eliminates uncontrolled light pollution.
3) It allows the pleasant use of space by people in the dark.
4) it achieves huge energy savings.
5) It motivates professional use of advanced control systems for LED lights, with limitless possibilities.

"Shadow is light's best friend" (author)

Figure 161: The Grand Canyon of the Colorado River is a mile deep. Photo: Julle Oksanen. Nature can provide great examples for the darkness designer for how to create a design canvas, starting with a background of total blackness and removing layers of darkness one at a time until the desired degree of glowing contrast is achieved.

Architectural outdoor lighting design, as a process, is also darkness and shadow design. The real role of the designer is to be a designer of darkness. Darkness fascinates us as human beings, but working with it successfully takes exceptional skills, open-mindedness and courage. Imagine a hypothetical situation in which the designer begins to create a lit environment with total darkness as the starting point. The darkness designer begins to remove, or eliminate, dark layers from the total black background, one layer at a time, until the desired degree of glowing light on the designed surface (e.g. on a facade) is achieved. This "shadow design" is a professional way to design nocturnal spaces. The professional darkness designer treats his/her task the way an artist treats the canvas, by painting gradations of darkness using light sparingly. A "whispering light" palette becomes the design tool.

For almost a century, in the era of technical lighting design, we have been producing glary, highly illuminated city structures from which the fascination of darkness is totally missing. The only way to succeed is to dismantle the old lighting installations and start from darkness. Some cities are in fact considering such lighting renovations in certain parts of the city infrastructure.

Studies of how we meet others in darkness play an important role when selecting the minimum lighting values required in Darkness Design. Certain lighting values are needed to meet the mandatory face-recognition distance for relaxed movement. With this tool, we can avoid creating environments that are too dark for comfort.

Figure 162: "Skyline - New York City, New York at night," by Trodel, is licensed under CC BY-SA 2.0.

Even in the busiest cities in the world, like New York City, nighttime fascination is based on shadows and darkness. We can sense all the fascinating elements of darkness in Trodel's photo. There is "physiological darkness" (light and shadow) + "psychological darkness" (emotions of love/hate) + "behavioral darkness" (darkness holds both attention and fear) + "biological darkness" (a sense of control over the darkness, with no fear of nocturnal animals) + "mental darkness" (darkness and death in a living city that never sleeps)+ "artistic darkness" (in the composition of lighting, does your mind hear the song "New York, New York"?).

I have yet to find any scientific article or psychology professional fully able to explain the scientific essence of darkness and its impact on human behavior. This observation reveals, I think, a clear link between design and the mystical essence of the "great unknown" as a factor linking semiotic connections in architecture and interesting design.

Cultural differences in darkness behavior will have a powerful influence on architectural Darkness Design metaphors and concepts. These differences cannot be explained by means of technical numerical light values. Lighting recommendations do differ somewhat from one country to another, but because all countries follow the CIE (transnational research organization) research results, cultural influence can't easily be identified in these standards. To understand the influence of culture on how we relate to and interact with darkness, we would need the help of anthropological studies.

Although my doctoral dissertation "Design Concepts in Architectural Outdoor Lighting Design, Based on Metaphors as a Heuristic Tool" was honored as the best doctoral thesis in the field of lighting, I did not take into account at the time all the impacts of light pollution on biodiversity. It was only after completing my doctorate and getting to know the research director of the Leibniz Institute, Dr Franz Hölker, that I learned of the worrying problems caused by light pollution. It was on that basis that I decided to pursue research activities into understanding the effects of light pollution on biodiversity. Studies done by Dr. Hölker, et al., gave me inspiration and motivation to create this book.

Earth Hour: a source of inspiration

Earth Hour is an annual coordinated mass effort worldwide to raise awareness of environmental issues focused on energy consumption as one driver of climate change. It is organized by the World Wildlife Fund (WWF) to raise awareness of climate change. The campaign calls on citizens to voluntarily limit or cease their electricity consumption, mainly lights, for a single hour on one day per year.

The first Earth Hour event was held in Sydney, Australia, on March 31, 2007, and it has spread to 188 countries around the world. Andrea Jechow, from the Leibniz Institute & German Centre for Geosciences, wrote an interesting article[33] about it that mainly deals with the effect of Earth Hour on light pollution in a Berlin city park.

However, the most important message is that people around the world are concerned about energy consumption and light pollution, even though few understand the real effects of it on biodiversity. This was the inspiration for this book and a motivation to do further research on the subject.

---

33 "Observing the Impact of WWF Earth Hour on Urban Light Pollution: A Case Study in Berlin 2018 Using Differential Photometry" https://www.mdpi.com/2071-1050/11/3/750

Japanese Shadow Design: An Inherent Protector of Biodiversity

The ancient art of shadow plays forms a natural cultural link to modern Japanese lighting design. Numerous similarities can be found between these two shadow design activities: reverence for shadows, telling stories with shadows, playing with directions of light, creating shadows instead of creating light, etc.

Famous Japanese author Junichiro Tanizaki wrote in his book *In Praise of Shadows*: "In making for ourselves a place to live, we first spread a parasol to throw a shadow on the earth, and in the pale light of the shadow we put together a house."

Figure 163: "Japanese traditional style SAMURAI house / 稲葉家下屋敷(いなばけしもやしき)," by TANAKA Juuyoh (田中十洋), is licensed under CC BY 2.0.

Figure 164: "Japanese traditional style SAMURAI house / 稲葉家下屋敷(いなばけ しもやしき)," by TANAKA Juuyoh (田中十洋), is licensed under CC BY 2.0.

Some typical Japanese shadow design elements

Skilled Japanese light and shadow designers, such as Kaoru Mende, who grew up in Japanese culture, naturally use many Japanese shadow design elements such as:

1) **Gradations of darkness** (exploring the realm between light and shadow). Instead of uniform use of light, using shadows that shift from top to bottom and from left to right. An infinite range of shadow gradation within light and dark is a unique Japanese design philosophy and technique.

2) **Contrast** (the dynamic interplay of shadow and light). Our senses are extraordinarily adaptable. We can handle 100000 lx sunlight perfectly, but 2 lx of moonlight might seem to us very bright as well. We cannot analyze contrast effects with our human senses as we can by measuring lighting values, like illumination and luminance. The best we can do is to apply Hopkinson's Scale, to avoid contrasts that are too strong or even produce glare, disturbing our visual perception. We have three fundamental and well-studied ratios for

differences in brightness (as measured with a luminance meter): 1:3 (for example in interior lighting/offices, so as not to tire the eyes), 1:5 (for example, used at building entrances to smooth the transition between peripheral areas and enclosed elements, such as entrance carpet) and 1:10 (for example, in a large display case with eye-catching products, as seen in department stores). These are constantly used illuminance values in daily lighting design. But contrast is rooted in actual visual perception, which must be taken into account in good Darkness Design. Colors, materials and space structuring are important elements.

3) **Layering** (the art of shadow upon shadow). Western cities are so overloaded with light that lighting designers have to use mainly "layered lighting," while Asian lighting designers have more opportunities to use the "layered shadow" philosophy in their design.

We have generally too much light in our illuminated landscape areas and indoors. We have to remember that "Shadow is light's best friend" and that shadow is born from light (Howard Brandston). We as human beings are intrigued by life's binaries—life & death, salt & sugar, love & hate, war & peace, light & shadow. We crave these contrasts. Consider old films like *Citizen Kane* or *Casablanca* or Charlie Chaplin's *The Great Dictator* or *The Third Man*, or dramatic moments featuring Marlon Brando in Francis Ford Coppola's *The Godfather*, which employ masterful use of shadow and light, accompanied by dramatic music. We need such contrast; we want it, but we are not skillful enough to handle it, yet. We need education and educated darkness designers.

Figure 165: "The Godfather – 1," by komersreal, is licensed under CC BY 2.0.

Shadow and darkness composition

We have long lacked a rational and comprehensive philosophy for creating beautiful nocturnal outdoor spaces. Understandably, there are many reasons for this failure to develop competence and tools for this over the past 100 years. But we do have good examples of how technical research can help to create a beautiful nocturnal ambience by creating a "lighting composition" for a certain architectural section of a city environment.

Enclosed light and shadow at Mariehamn Harbor

For this project in Finland, a fine lighting, shadow, and darkness composition results from an intriguing and well-tested heuristic metaphor. In the innovation phase, we were looking for a "heuristically analyzed lighting design metaphor element."

What we found could be summed up in several phrases: music & light; Richard Kelly's design process; "shadow is light's best friend"; contrast of black sea and soft white light veil; poetic light. The composition of light and darkness as depicted in the architectural section drawing does not follow scientific method, but this technique has proven to be a great tool for architectural lighting and Darkness Design.

It is advisable to divide a project area into well-thought-out parts, each with its own architectural and darkness/lighting composition. This can be thought of as corresponding to music. Just as in music, the pause is as important as the note; in biodiversity-friendly lighting, the composition of shadow is just as, if not more, important than light. (Shadow is light's best friend.) Decisions about lighting a particular section can also take into account light technical values, such as a semi-cylindrical minimum illuminance value of 0.8lx.

A lighting section composition can actually be sung or played to the client on a musical instrument: the higher the lighting value, the higher the pitch. Reading from left to right in the Mariehamn Harbor design, the "music" proceeds from the ship *Pommern* to the Sea Promenade, the Museum and finally the harbor.

Of course, the overall solution must follow the ULOR-CT principles described earlier.

Concept:

Before composing sections of light and darkness, it is important to create a clear lighting design concept for the project. The poetic concept for the Mariehamn Harbor could be described like this:

When darkness falls on the Mariehamn area, a veil of nocturnal light pervades the whole space, forming a "glowing harbor." The white veil over the harbor contrasts with the mystical, murky sea, a dichotomy of life. *Pommern*, the world's biggest four-mast barque, floats dormant in the glowing white mist. The captain is walking alone in his captain's cap and uniform, a brown leather briefcase in one hand and a pipe between his lips. He is walking slowly in the glowing white light towards his home, the smoke from his pipe trailing behind him, slowly dissipating. The harbor falls into a deep sleep.

Electric light is composed to recreate this nocturnal scene, which is more or less a dream, a personal vivid memory from Orson Welles' "The Third Man."

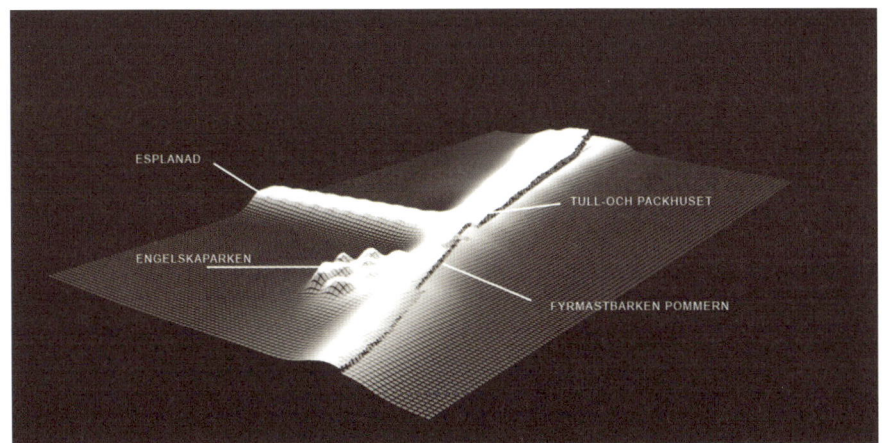

Figure 166: Mariehamn Harbor veil philosophy. Photo: Julle Oksanen and Oliver Walter.

Figure 167: Lighting sections 1,2,3 and 4. Photo: Julle Oksanen and Oliver Walter.

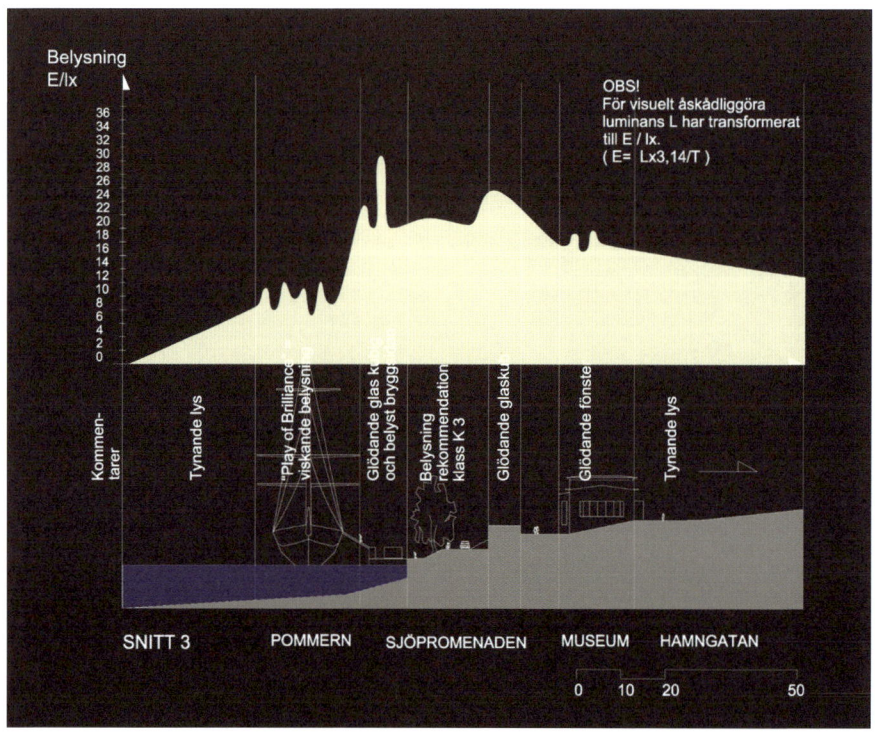

Figure 168: Lighting section number 3. Photo: Julle Oksanen and Oliver Walter.

Light and shadow composition for section 3. The vertical element in the diagram is E, the illuminance value in lx. The yellow areas represent carefully designed lighting compositions. The basic lighting "tone" is a few luxes (3-4 lx) over the whole lighting section. That represents the basic "white light veil." The yellow area above is approximately 3-4 lx basic tone and the various bulges represent "focal glow" lighting (like the light cubes of the new museum building) and "play of brilliance" lighting (like the illuminated masts of the *Pommern*). A Brando luminaire is located on the left side of the Sjöpromenaden. It is not exactly on this section, but causes approx. Ehor = 18lx on the street surface (Brando pole distances selected according to light traffic classification K3). On the left edge of the section, the light is fading to the sea; on the right edge of the lighting section, it fades into the park area.

Figure 169: The *Pommern* and lighting principle in section 3. Photo: Julle Oksanen and Oliver Walter

New technical tools for designing darkness

In Western culture we have over-indulged in nocturnal lighting to the point that in urban environments we must impose light over light to bring out structural elements. Eliminating light pollution requires a less heavy touch. There is an effective cure: professionally coordinated education in nocturnal design for everyone involved in such projects. By considering the structures of twilight and darkness, the designer discovers elements such as gradations between light and dark, the dynamic interplay of light and shadow, the art of shadow upon shadow, finding the rhythm of darkness, etc.

In the hype over LED lighting, the main objective—to reduce energy consumption—has already been lost. An even more annoying failure is the preponderance of LED units that produce tremendous glare. The engineering-dominated perspective that focused on efficiency and sharp aim, produced by the trend toward small and efficient light-producing units, has created difficult glare conditions. In the following sections, we'll take a look at whether intelligent lighting control systems and, for example, the use of Hopkinson's Scale can correct this situation and move us towards a tolerable Japanese-style perception of nocturnal spaces. This issue underscores the need for better training.

Smart Cities and lighting control systems

Modern lighting control systems are an amazing tool for cutting energy consumption while creating comfortable Darkness Design on a broad scale. An example of what can be achieved is the Interact City Lighting system, from Signify:

Lighting asset management

This software gives operators full visibility into the lighting infrastructure. Automatic fault detection alerts them to issues needing a fast response, with minimal downtime. Data can be used to make informed decisions and optimize lighting performance. It makes it possible to manage lighting-related workflows from an intuitive application and view data from a centralized dashboard.

Energy optimization

Control systems allow the city to optimize street lighting performance and measure energy usage accurately in real-time. Full control of city lighting makes it possible to reduce CO_2 emissions, make progress toward sustainability goals, and lower energy consumption and costs. These savings can be reinvested in other areas of the city infrastructure.

Scene management

This feature makes it possible to remotely adapt city lighting to suit the time of night, season, or event. Lighting can be turned up if there's a traffic incident or a crime in a particular area, or dimmed to 30% when the streets are empty late at night. Sensors on light poles can be used to detect activity, keeping citizens safe and comfortable and easily turning parks and plazas into livable spaces.

Sensors

These can turn every street light into a city sentinel. Outdoor sensors that detect motion/presence, tilt, vibration, ambient temperature, noise and other factors can be attached to a luminaire fitted with the ZHAGA Book 18 push-and-twist lock socket interface. The sensing functions can be configured remotely, and data can also be sent directly to the Interact City application.

Smart street lighting is part of the smart city environment.

Integrating smart street lighting into a central dashboard enables the lighting system to communicate with other smart city applications, such as smart parking, waste management, and traffic control. How does it work? Through this integration, the customer is able to extract, analyze, and utilize the data generated from various systems such as transportation, environment or traffic. This benefits all stakeholders across the whole range of municipal services.

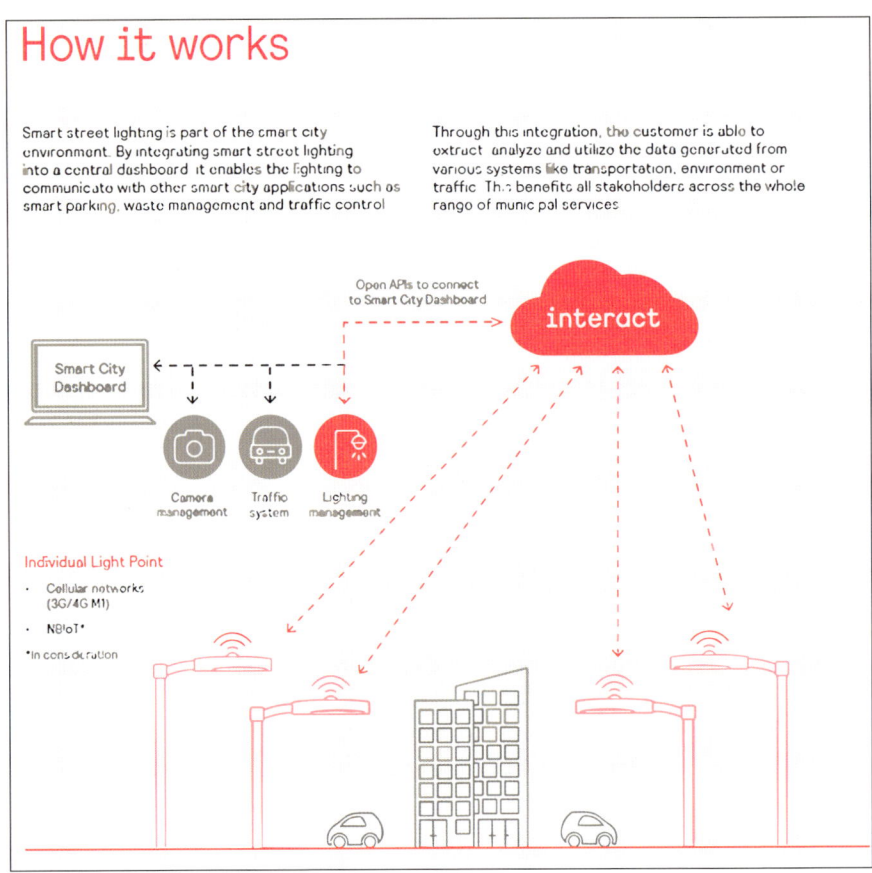

Figure 170: Integrating smart street lighting into a central dashboard, as part of smart city management, enables the lighting to communicate with other applications such as smart parking, waste management, and traffic control. This integration allows the city to analyze and use data generated from various municipal services, such as public transportation, environment sensors or traffic data. Interaction between buildings and other elements of the urban infrastructure is missing from this picture because of the complications of ownership and various obligations, but such barriers may be overcome in the future.

Integrating facade lighting into urban intelligent control systems

A smart urban lighting system is only a tool for professional Darkness Design, not a solution for the underlying issue of protecting biodiversity. But it points us toward a new model for lighting public buildings.

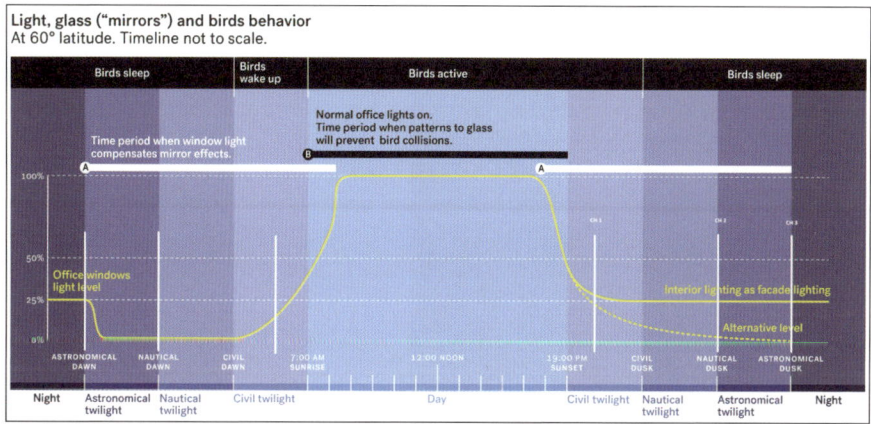

Figure 171: Integrating interior-lighting-as-facade lighting into, say, the central dashboard of the Smart City Lighting System, would enable this biodiversity-friendly lighting solution to communicate with other smart city lighting applications. Through this integration, the city could get "whispering lights" facades in every public building. (In the Lights, Glass and Birds diagram, the yellow line describes the controlled lighting level during the day and after normal office hours: Channels 1-2-3, etc.)

This would eliminate light pollution in the sky, prevent bird collisions and deaths at the windows, and create pleasant nocturnal design solutions in the urban architecture.

At the moment, the hierarchy of the city's built environment clearly distinguishes between lighting belonging to the city and that of private buildings. Building owners have traditionally taken care of their own lighting. But this could become a joint effort to eliminate light pollution and achieve energy savings of as much as 80%, through a unified "City, People and Light" philosophy. Developing this overall system would require design cooperation among biologists, architects, engineers and urban planners.

Eliminating light pollution from nocturnal architecture requires, in the author's opinion, sound and holistic knowledge of "old architectural lighting design." A good tool for this is the author's doctoral dissertation "Design Concepts in Architectural Outdoor Lighting Design, Based on Metaphors as a Heuristic Tool," which can be downloaded at https://aaltodoc.aalto.fi/handle/123456789/27886. This book is a follow-up to this dissertation, which is highly recommended for readers who want to fully understand the principles of nocturnal Darkness Design and its efficacy for eliminating light pollution and its effects on biodiversity.

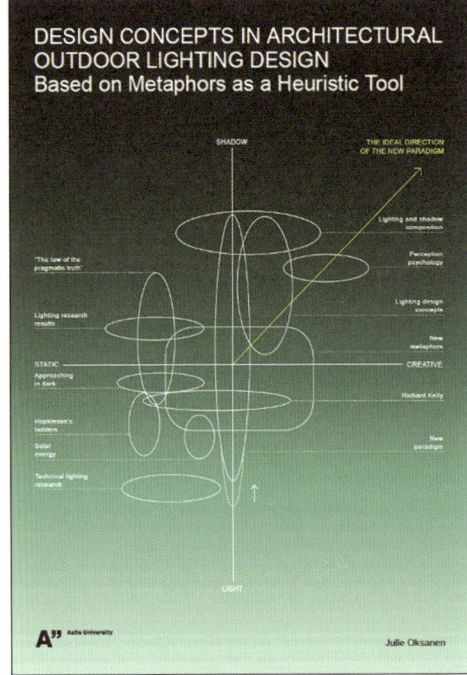

Figure 172: https://aaltodoc.aalto.fi/handle/123456789/27886

NOCTURNAL MASTER PLAN STRATEGIES FOR CITIES

"Shadow is light's best friend"

Architectural outdoor lighting design, as a process, is simultaneously darkness and shadow design. To reiterate, designing darkness in a fascinating way takes exceptional skill, an open mind and fearlessness.

The darkness designer (formerly lighting designer) begins to design a lit environment with total darkness as the starting point. The designer removes dark layers one at the time, until the desired lighting degree on the designed surface (e.g. on a facade) is achieved. Thus, *lighting* design becomes *shadow* design.

As described earlier, shadow design can be compared to oil painting techniques, the way Leonardo da Vinci used successive ultrathin layers of paint and glaze to give his works their dreamy quality. The darkness designer treats his/her task like an artist, painting gradations of darkness on the canvas using light sparingly on illuminated surfaces, thereby employing a "whispering light" palette as the design tool.

Darkness Design is a challenge within the context of the glaring and over-illuminated city environments that have been developed over the past 100 years, where the fascination of darkness is missing. Success will mean dismantling old lighting installations and starting from darkness.

Eliminating light pollution from nocturnal architecture requires advanced co-operation and a good understanding of the overall situation from all parties involved. It is wise to deploy modern technologies, for example advanced "smart city" interactive systems and skilled experts in intelligent control of LED lighting. Since smart city lighting uses central computing and comprehensive control, it makes sense to carefully examine the city's overall lighting situation before the system is introduced, partially or comprehensively. Overall strategies for eliminating light pollution and moving smoothly forward can then be created. As a first step, it would be good to analyze the entire illuminated structure of the city, although naturally not everything will be rebuilt immediately.

Step 1: Analyzing existing lighting

Figure 173: An example of analyzing existing lighting, from the report "Lighting Master Strategy for the city of North Vancouver." Photo: Lighting designers Gabriel Design & Tania Sagoo Lighting design.

For a city, developing a light pollution–free nocturnal Darkness Design master strategy has many important biodiversity-friendly advantages. It brings together a wide variety of city authorities, from people in the planning office and the city construction department, to city architects, computerization experts, electricity designers, biologists, leaders of building management, and in the later phases electrical utility contractors, and so on.

Cities can save money by working collaboratively on the same area (e.g., installing electrical wiring in tandem with water pipes, etc.), once everyone knows how parts of the city area will be covered in the new smart lighting/interior-lighting-as façade-lighting program. Cooperation between design professionals—architects, electrical engineers, darkness designers, smart city system architects, etc.—can be planned better.

Analyzing the issues together would create an opportunity to start working seriously with different kinds of nocturnal metaphors and using heuristics on a large scale. (For example, "whispering" interior lights for facade lighting; shadow design for parks, water and natural areas; various white color temperatures; eliminating glare from various kinds of streets, etc.). The biodiversity-protecting nocturnal design solutions discussed in the book, together with smart city lighting methods, provide opportunities for bold design teams to shift their overall lighting strategy in urban centers toward enjoyable nocturnal design.

Step 2: Character zones and defining the testing area

It's important to understand the characteristics of different urban areas. In the picture of the example city (North Vancouver, Canada), the gold-colored area is a bustling downtown, the darker red areas are quieter neighborhoods, and the green areas are park areas in the city.

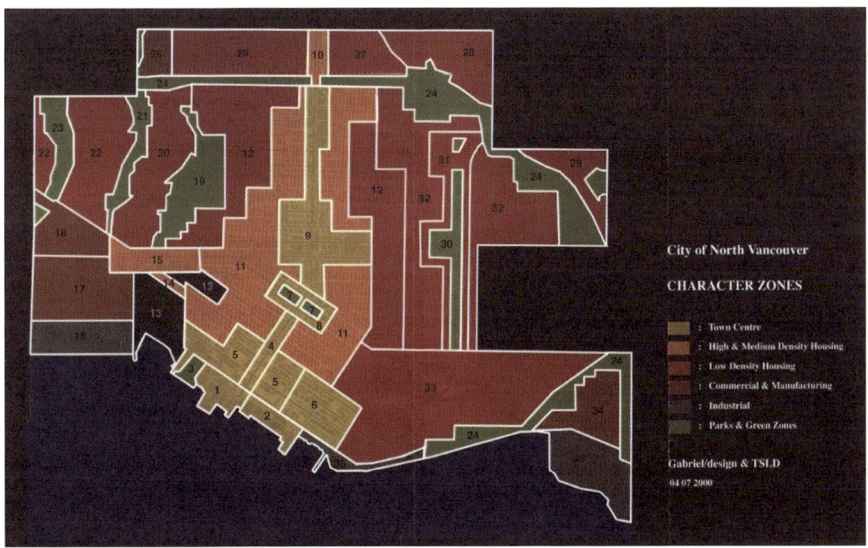

Figure 174: Character zones of North Vancouver. Photo: Lighting designers Gabriel Design & Tania Sagoo Lighting design.

The appearance and zoning of the city's urban structures suggests that certain areas be given priority. This facilitates light-pollution studies that will be carried out later, for example using drones. Zoning determines where the project work should begin and where more limited tests can be done, and so forth. Nocturnal design elements of the Smart City Interact lighting program and the devices and lighting systems to be integrated into it are best implemented by first selecting a suitable test area. That way, test results can be viewed and, if necessary, faults can be corrected quickly without incurring high costs, while putting in place, for example, various equipment functions, lighting control, biodiversity-friendly interior lighting as façade lighting, and light pollution control.

Before implementing the system in the test area, it will be necessary to ascertain the routes for different modes of mobility (pedestrians, bicycles, cars, public transport, etc.) as well as to compile a list of the city's interesting objects for potential night illumination. Nocturnal Darkness Design for those routes and objects, free from light pollution, must be considered before going on to design and implement the pilot project. Some Darkness Design proposals will be introduced at this stage, following research on biodiversity-friendly measures.

Step 3: Circulation and the testing area

The basis for classifying different kinds of traffic routes (for pedestrians, bicycles, motorcycles, cars, and public transport, as well as heavy traffic) as to appropriate lighting was previously grounded in the use of un-adjustable gas discharge lamps. Existing lighting has been based purely or mainly on technical recommendations. However, the old lighting has since been replaced with LED units that can be adjusted from 0 to 100%. This calls for a reconsideration of these classifications and recommendations.

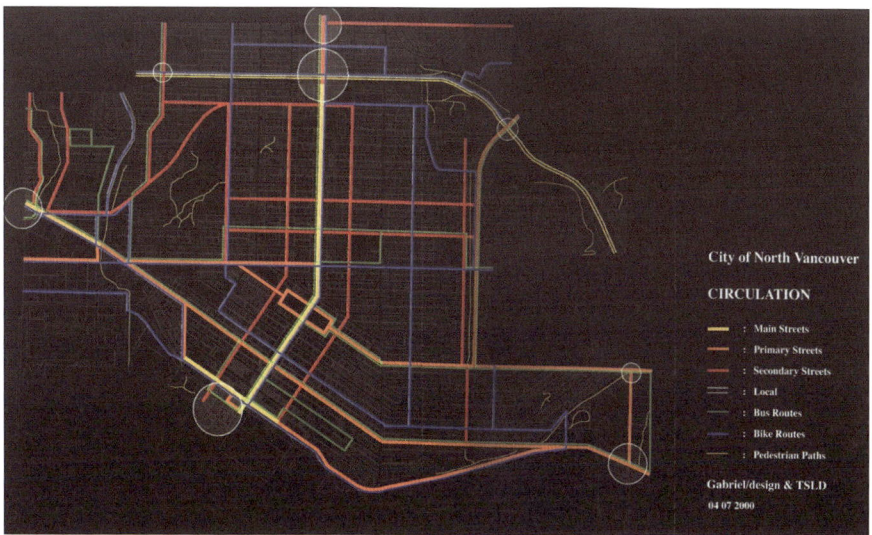

Figure 175: Traffic in North Vancouver. Photo: Lighting designers Gabriel Design & Tania Sagoo Lighting design.

Space, Darkness Design, traffic LED lighting & glare

The advent of LED lighting has largely failed to produce a soothing and stylish twilight architecture, or illumination that supports the diversity of nature. It is physically impossible to produce pleasant light with very small light output surfaces and high luminance values, even though suppliers have invested in ensuring precise alignment capabilities in LED units. To visualize why this so, imagine two spheres of light, each producing the exact same amount of light, but one a meter in diameter and other just 5 centimeters. Evaluating existing lighting and comparing what might come from a new comprehensive solution will certainly affect what will be applied to the test area, depending on its size and a heuristic cost estimation.

LED luminaires and glare ratings

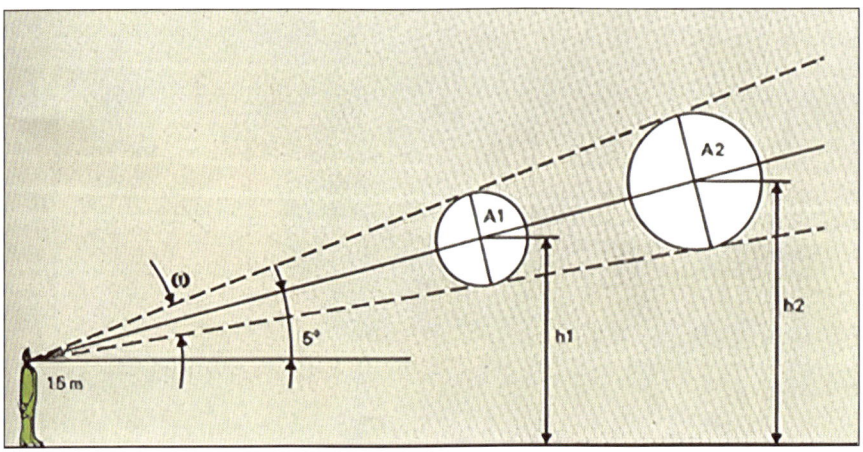

Figure 176: Principal drawing for LA 0.25 value. Photo: (Philips ILR 1980 /3)

If the surface of the luminaire is brighter than the recommendation allows, it affects the distance from which we are able to recognize others as we approach them at night. That's because glare veils the retina and makes it harder to distinguish contrasts. If recognition distance is shorter than 10 feet, we may feel uneasy and perceive the situation as unsafe; chances are good that at least one person will change their walking direction. A correct "recognition zone" is valuable for ensuring relaxed and enjoyable movement in darkness. The luminaire glare value calculation formula is:

LA 0.25
where,
L = measured luminance between angles of 85-90 from vertical (bright area section of the luminaire light distribution surface),
A = bright luminaire light distribution section.
The maximum values depend on the height of the luminaire:
$LA^{0.25}$ = 1250, h < 4.5m
$LA^{0.25}$ = 1500, 4.5m < h < 6m
$LA^{0.25}$ = 2000, h > 6m.

Calculating the glare value of a randomly selected LED traffic luminaire

Depending on the approach direction, using a randomly selected small-road lighting type meant for parks, a calculation of glare value rises to **LA 0.25 = 9400**. The light output value of the luminaire must be reduced by 84% if it is to meet the acceptable value for light-traffic routes, for example, mounted on a pole 5 m high, with maximum **LA 0,25 = 1500**. For the nocturnal architecture design team, designing this is a demanding but not insurmountable task. The luminance value of this same light-traffic luminaire in the direction C = 150 degrees and Gamma 65 degrees is more than 200,000 cd /m2. That can be described as foolish. The glare of different kinds of existing luminaire types on the area can be calculated by measuring the luminance (L) of each luminaire type with a luminance meter, as shown in the main picture, and calculating the projection areas (A) of different luminaire types.

Blinded by the light

A recent study by the Emergency Responder Safety Institute in the United States revealed that the LED flashers on police cruisers, fire engines, and ambulances that spray the night with intense beams to protect first responders from oncoming traffic actually create a lot of glare, and that effect intensifies when the light bounces off retroreflective chevrons on the backs of emergency vehicles. The study found that substantial reductions in intensity did not make the lights any more difficult to detect along the road.

Calling their findings (particularly about the effects on the high level of retroreflectivity) "surprising," the researchers determined that rubbernecking motorists can be blinded by intense LED lights and fail to see emergency personnel on the roadway.

The US Department of Transportation reports that vehicle accidents are the second most common cause of fatalities among police officers and firefighters, and the leading cause of death among tow truck operators.

One of every five of those deaths can be tied to a secondary accident, such as when a motorist squinting through the glare of lights fails to see an emergency worker before it's too late.

Figure 177: Photo: Emergency Responder Safety Institute.

A partial solution in the testing area: intelligent LED headlights

As all car manufacturers are engaged in development of smart LED headlights, it would be important to test minimizing road lighting values, switching to very low or even turning them off altogether in testing areas. The overall concept must be considered (including the light output to outdoor space of interior lighting in buildings, the overall effect of park lighting on the space, etc.). An interesting test could involve dimming road lighting a little more each week. It would be good to publish information about the test area to the local media to avoid unnecessary grumbling.

Mercedes Benz as an example

In the new Mercedes Benz E-Class, the optional high-resolution Multibeam LED headlamps, each with 84 individually controlled high-performance LEDs, automatically illuminate the road with a precisely controlled distribution of exceptionally bright light—without dazzling other road users. That's because the grid design allows the light distribution of the left and right headlamps to be controlled separately and adapted quickly and dynamically to the changing situation on the road. All functions of the Intelligent Light System in low-beam and high-beam mode are activated digitally, without mechanical actuators.

 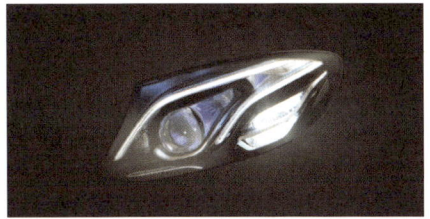

Figure 178 left and right: Multibeam LED headlights, three-stage precision optical system. Photo: Mercedes-Benz E-Class: Multibeam LED

Figure 179: For perfect visibility, the adaptive Multibeam LED headlights with individually controlled LED lights respond to the traffic situation. Photo: Mercedes-Benz E-Class: MULTIBEAM LED

Employing Hopkinson's Scale in the testing area

When all traffic routes are under control (dimmed to the minimum or completely turned off), the nocturnal architectural space can be shaped fantastically, according to a new paradigm for nocturnal space. The tool for this is Hopkinson's Scale.

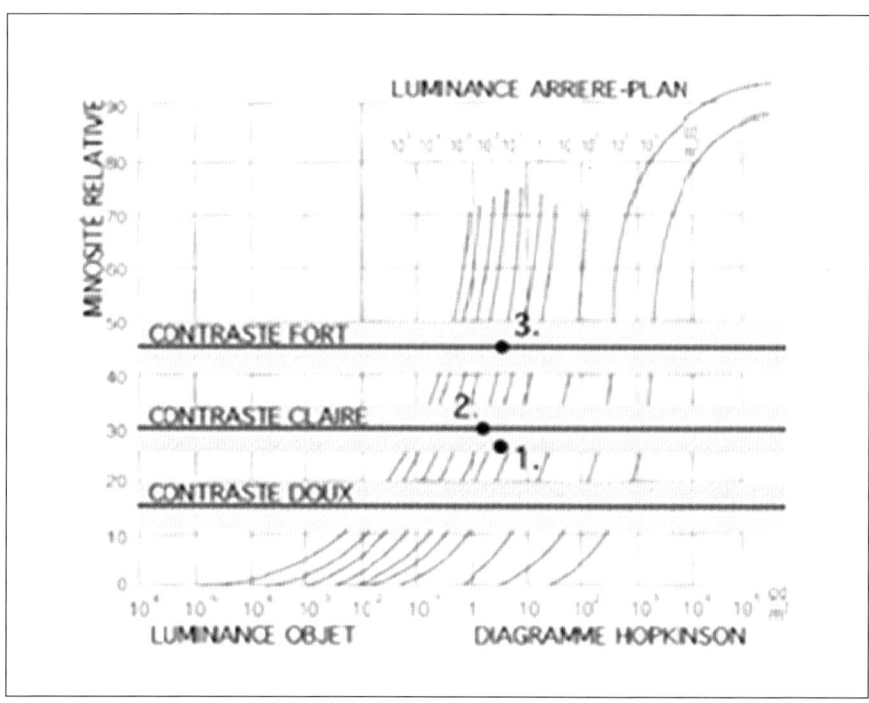

Figure 180: Biodiversity-friendly and light pollution–free contrast levels for outdoor Darkness Design are described in this example from a project in Arras, France, as: "Contraste Doux = sensitive contrast, "Contraste Claire" = clear contrast, "Contraste Fort" = high contrast. Photo: Julle Oksanen's dissertation. Shadow design example of the Cathedral in Arras. "Sensitive contrast" and "clear contrast" are recommended for nocturnal spaces. Use a luminance meter and mark values on the test area design map.

LED solutions in the test area: large-surface lighting

In the author's opinion—and perhaps it's a collective opinion—lighting across a large surface produces a calming and harmonious effect. Glare is absent from large light-distribution surfaces in an LED luminaire, meaning that it doesn't create veil luminance on the retina of the human eye. The space is peaceful, the contrasts are clear, and it is easy to see details. Variations of shadow are easier to achieve than in a situation where numerous post top fixtures and/or road lighting fixtures "shoot" light beams into observers' eyes. A large-surface lighting solution creates an elegant ambient light for the space. It is enjoyable to design "focal glow" and "play of brilliants" elements in this kind of environment with a small amount of light. Well-designed large-surface lighting can totally eliminate the need for separate road and street lighting. Nocturnal city beautification is easier to achieve without big, ugly, glaring light fixtures atop massively high columns.

And it's possible to achieve huge energy savings. By integrating solar panels above the LED panels on the structure, it's possible to produce remarkable amounts of free and clean solar energy for our use. Payback time for the investment can be easily calculated thanks to the energy savings derived from this practice.

A biomimetic metaphor design process for large-surface lighting is introduced in Appendix 2: "Futuristic City 2030". The theory can be summed up in a few phrases: "a heuristically analyzed biodiversity-friendly lighting design metaphor"; a city without poles and ugly luminaires; sustainability using solar energy; biomimetic LED panels integrated into building structures; Richard Kelly's "ambient light," "focal glow" and "play of brilliants"; a totally glare free solution; Hopkinson's Scale; full return on investment in a few years.

Figure 181: Large-surface lighting. "City 2030." Design Julle Oksanen. Photo: Oliver Walter and Dan Silberman

In the testing area, it would be useful to introduce and test self-made, adjustable large-surface luminaires by temporarily installing them in the space. The appropriate size of the light panels could be similar to the panel of the Brando luminaire, atop a 6- to 8-meter pole. Either a one-side or two-sided Brando, depending on design context.

Experimental adjustable LED light panels, as shown in the picture, meant to be installed on wall surfaces, would make an excellent research task for a student group that includes a biologist, engineer and architect.

Appendix 2, "Futuristic City 2030," presents a more detailed and comprehensive structure for the research process of an interdisciplinary student group of this type.

Step 4: Places of night interest in the testing area

Intriguing elements that pique the interest of passersby can be found in many urban structures: important buildings, statues, trees, water elements, etc. Any sites destined for special treatment should be selected by a professional team that includes urban architects, engineers and, hopefully, also biologists, who could contribute their insights on biodiversity issues (bird migration routes and high buildings, park lighting, water status, etc.) early in the process. When selecting objects, it is paramount to remember the importance of protecting biodiversity, as well as the need to create nocturnal solutions free of light pollution.

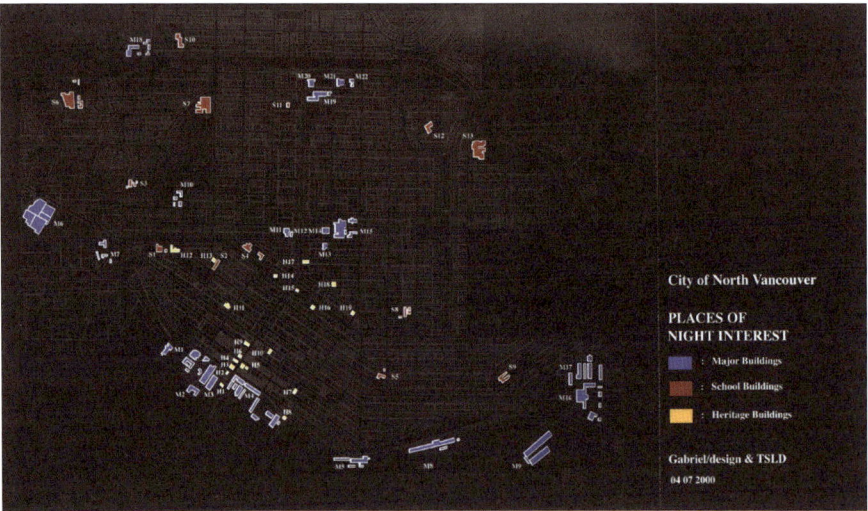

Figure 183: This plan for North Vancouver shows places that will benefit from night lighting that recognizes certain landmarks. Lit buildings at night help with orientation and add a feeling of safe passage. Photo: Lighting designers Gabriel Design & Tania Sagoo Lighting design.

If façade lighting has not yet been integrated into a centralized control system, it is good to choose office buildings where it is possible to have the control system communicate with the building's internal systems, following the curve.

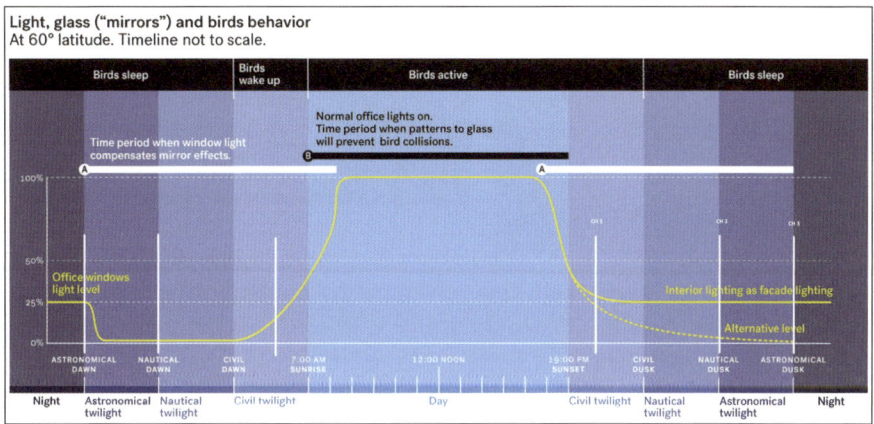

Figure 184: Integrating "Interior Lighting as Facade Lighting" (yellow line starting control lighting levels after normal office hours Channels 1-2-3-etc.) into for example the central dashboard of the Smart City Lighting System would enable the biodiversity friendly lighting solution to communicate with other smart city lighting. If this façade lighting system has not yet been integrated into centralized control systems (manufacturers development work in progress), it is good to choose office buildings where the control system can be programmed into the internal systems of the selected buildings according to the curve.

Don't test façade lighting

It takes high light output to illuminate large facade surfaces. When a building is illuminated this way, the large amount of ambient city light already present will need to be taken into account when choosing floodlights. In the picture below, façade lighting on a high skyscraper has been accomplished using almost a hundred high-power floodlights. The effect of this on light pollution is shown in the figure on the right. Huge amounts of direct and reflected light from façade surfaces (windows and solid material) illuminate space in vain.

Figure 185 left and right: Left: One skyscraper façade lighting luminaires and light pollution producers in Times Square, NYC. Photo: Julle Oksanen. Right: "Times Square and Broadway at night from the Empire State Building" by mattk1979 is licensed under CC BY-SA 2.0

Heuristic cost calculation for the skyscraper: Power / 60 units x 1000 W/unit x 4 facades/building = 240 kW / Annual use 4000 hours / Electricity cost 9.43 cents/kWh (0.1€/kWh) / Annual electricity cost 96,000 € + approximate annual maintenance cost (lamp + work) 15,000 €. Total annual costs 111,000 €.

Total annual costs if LED luminaires are used instead of metal halide, but with same lumen output: 96,000 €.

Do test: interior lighting as facade lighting

Unlike bright, large-facade lighting solutions, relying on the building's adjustable interior lighting to highlight the building creates a cozy and inviting atmosphere in the urban area. Energy savings are huge compared to floodlight installations—up to 96%, according to heuristic calculations. This also means there is no unnecessary illumination of space and sky in the test area.

It's a good idea to perform Hopkinson's Scale measurements in the test area. From the background measurement area of the field of view (including the facade, trees, sky, etc.) it is easy to measure the degree of brightness of the background with a luminance meter. The luminance of the building is then the luminance value measured from the windows.

Using this type of lighting, urban buildings don't compete with each other for attention but create a holistic harmony within the urban space, keeping the skyline in harmony, too. As outlined above, the facades of some important buildings may require separate floodlighting, but only those so considered by people responsible for the city's nocturnal design and designated in the Strategy Master Plan. All illuminated windows in large public buildings in an urban structure, their lighting calibrated on a "Light-Glass-Bird" basis, produce the vertical light of the urban space and are a significant part of meeting the overall "whispering lights" principle.

A new operating model that protects biodiversity might make the city responsible for interior lighting bills in privately owned buildings, in return for the building owners' agreement to cooperate with the master plan and a control system that makes interior lighting serve as facade lighting.

Figures 186 left and right: Left: Quiet architectural "whispering" light in stylish New York City. Photo: Mauricio Chavez on Unplash. Right: Telenor Building, awarded the Norsk Lyspris Prize in 2003. Metaphor: House as a luminaire, interior lighting as both façade lighting and plaza outdoor lighting. Photo: Jan Drablos

Other interesting elements in the testing area

It is important that the ULOR-CT value (Upward Light Output Ratio-City Total) resulting from the entire test area is 0. However, the lighting hierarchy outlined by Richard Kelly is sometimes needed to maintain interesting spaces and create comfort, examples being the Gateshead Millennium Bridge and "Lighting The Lady."

Figure 187: Bridge lighting example: Light coves and ULOR-CT = 0. Photo and details: Julle Oksanen and Oliver Walter

Figure 188: Glare-free, peaceful, "whispering light" on a small plaza and pedestrian bridge over the Tennessee River generates a quietly romantic ambience for pedestrians who cross here daily. Project ULOR-CT = 0. (Julle Oksanen's studio work at the University of Tennessee. Photo: Larwie and Associates, CRJA, Wilbur Smith Associates, S&ME, Sanders Pace Architecture, Julle Oksanen Lighting Design Ltd., Julle Oksanen & Oliver Walter.)

 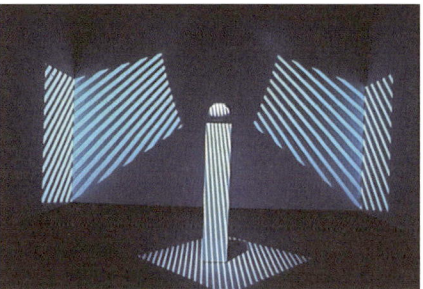

Figure 189 left and right: Bridge lighting example: Left: "play of brilliants" case study, Gateshead Millennium Bridge. Lighting design Speirs + Major. Photo Andrew Curtis, licensed under CC BY SA 2.0. Right: "Play of brilliants," theoretical display. Photo: ERCO/Julle Oksanen's dissertation

Figure 140: Focal Glow, case study Howard M. Brandston Lighting The Lady. Photo: Brandston Partnership inc.

The metaphor of "Lighting the Lady" offers a clear example of using heuristics in creating design. "I pondered the challenge of how to illuminate this icon [the Statue of Liberty in New York], so I took a boat out into the harbor and observed her from various distances and at every angle," Howard Brandston wrote of this project. "I observed her at dawn, noon, dusk, and in the darkness of night. At some point of the process, I walked to the end of the Battery Park Promenade, sat on a stone wall, and it came to me: She looked best in the light of dawn. I took out my little notepad and wrote: "Needed: one light source with spectral power distribution to mimic the morning sun, one to mimic the morning sky, and a new light fixture to project light from a great distance—this creates a lady with green skin looking good." (Brandston, H. 2008). Photos: Howard Brandston/Julle Oksanen's dissertation

Step 5: After testing, implementation

Based on experiments performed in the test area, the methods obtained from the lights and shadows as well as the visual and lighting numerical values can be successfully combined with the city's Master Plan.

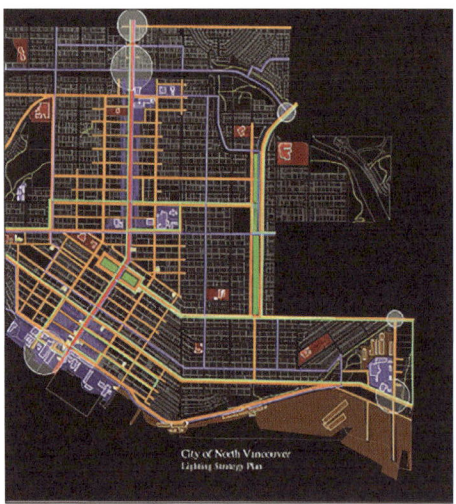 

Figure 191 left and right: Left: Example of the Lighting Master Plan of the City of North Vancouver. Right: New street lighting proposal of the City of Vancouver. Photos: Lighting designers Gabriel Design & Tania Sagoo Lighting design.

OUR FUTURE

Finding real solutions to the problem of light pollution and its effects on biodiversity is a huge but important global project, and in a time of rapid climate change and other pressures on our natural environment, there is no time to waste. It will take a high level of leadership among international stakeholders and the cooperation of people in many fields, with responsibilities allocated through the appropriate channels, to find agreement and take effective action.

This book has suggested some of the organizational changes and new concepts in nocturnal architectural design needed to get serious about eliminating light pollution and protecting the natural world. Actions must be taken decisively, cooperatively, and logically, because new technologies and proposals are moving forward so quickly.

We can only make this strategic shift together. This is my dream.

I send faith, hope and love to all who come across this book. Thank you for reading it.

Julle Oksanen

APPENDIX 1: TELENOR: ORCHESTRATED DARKNESS DESIGN IN A HIGH-RISE BUILDING

We carried out this project in 1998 completely unaware of the destructive effects of electric light on biodiversity. Architect Vesa Honkonen's natural ability to integrate shadows into his architecture created a good basis for implementing this project. Each time the words "Lighting Design" appear in this appendix, they can be replaced by the words "Darkness Design," the future path of biodiversity-friendly design.

Any professional architectural lighting design follows a basic four-step method, regardless of the project type. These steps consist of:

a) Heuristic metaphor & concept design.
b) Master Plan design.
c) Detail design.
D) Demonstrations and implementation.

This sounds simple, but carrying it out requires a professional attitude, skills and education. There is no exact, standardized design process inside these four important steps, because the method must be based on the specific setting of the project to achieve quality results. Every project needs setting-based creativity and project management. The Telenor headquarters lighting design project is an outstandingly comprehensive and illustrative example of this.

Figure 192: Concept example. The Telenor building in Oslo, Norway. Photo: Telenor

Telenor is one of the fastest growing providers of mobile communications services worldwide and the largest provider of TV services in the Nordic region.
- Revenues 2006: NOK 91.1 billion
- Workforce: 33,500 person-years
- Listed on the Oslo Stock Exchange and headquartered in Norway.
- The company was looking for a lighting design concept for its new headquarters in Oslo. The building incorporated a lot of glass. Focusing on those glass areas and the plaza was our design task.

Darkness Design concept

Figure 193: Telenor aerial view. Photo: Telenor

This office building is near a beautiful fjord, in a natural setting. As lighting designers, we saw Norwegians as people of nature. When we started creating the concept, we had in our minds an image of Norwegians wandering around on their mountains and fjords wearing the famous, characteristic Norwegian pullovers and rucksacks.

However, we kept in mind that Norway is also very wealthy, with a lot of oil and great educational opportunities. It is a modern nation, of which this Telenor building is a good example. This was somehow an odd combination to consider in the design phase. Seven thousand employees enter this modern, high-tech building daily. They approach the main entrances through a huge plaza that's about the size of four soccer fields, entering through eight main gateways located on the oval glass perimeters of the two building blocks.

Once through these, they arrive at an indoor entrance made mainly made of glass, steel and stone. After this they proceed to their office desks using stairs or elevators.

The next figure of the concept design shows a section of the office entrance hall block, located between the office blocks and made mainly of glass. A café is located in the hall.

Heuristic metaphors

We started the process of creating a metaphor by thinking about what Norway is. Norway has a lot of darkness and beautiful and wild nature, as well as Northern Lights and stars. We wanted to bring the darkness from the fjord to the site, floating silently into the building through the eight glass entrances, flowing through the building back to freedom outside. We wanted to use the Telenor buildings as huge luminaires on the plaza, their glowing facades forming "whispering lights." Our dream was that when darkness engulfs the site and pervades the entrances of the buildings, the glass facades would in turn glow modestly, shedding vertical light onto the plaza. We wanted no lighting fixtures at all on the plaza. In this design, indoor lighting would serve as outdoor lighting, producing a pleasant pairing of light and dark.

Based on that metaphorical poetry, our concept was to create a place for employees where they could have a clear and visible connection to nature. We imagined a person in the café looking upward to see the stars and moon in the Nordic sky.

Glass acts like a mirror. The reflection factor of normal "float" glass is something like 20%. The hall block was full of revolving glass elements set at different angles, both vertically and horizontally. We therefore decided to use a "functional lighting concept," meaning that we would try to minimize reflections from the glass surfaces. These entrance halls were actually "light traps" during dark periods. We used large surfaces and placed lighting units in such a way as to avoid or at least minimize reflections.

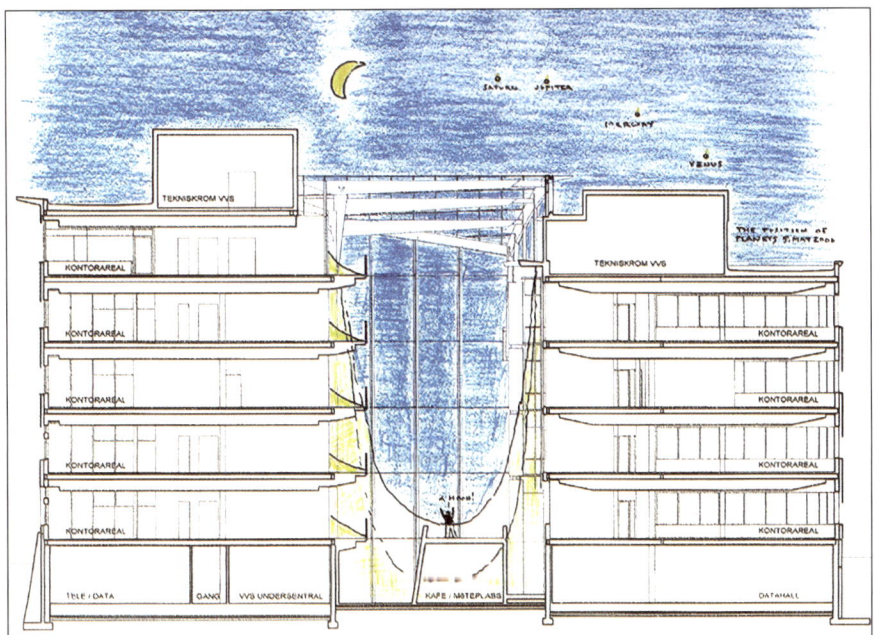

Figure 194: Telenor lighting design concept. Atrium section. Employees can see a Northern sky and stars from the café. There is no glare and no disturbing light surfaces to be seen. Photo: Vesa Honkonen & Julle Oksanen.

It would have been easy to integrate downward lights into the beams in the roofing area, a common solution. But we did not want to illuminate empty air, and we estimated that there would have been hundreds of reflected light sources as result.

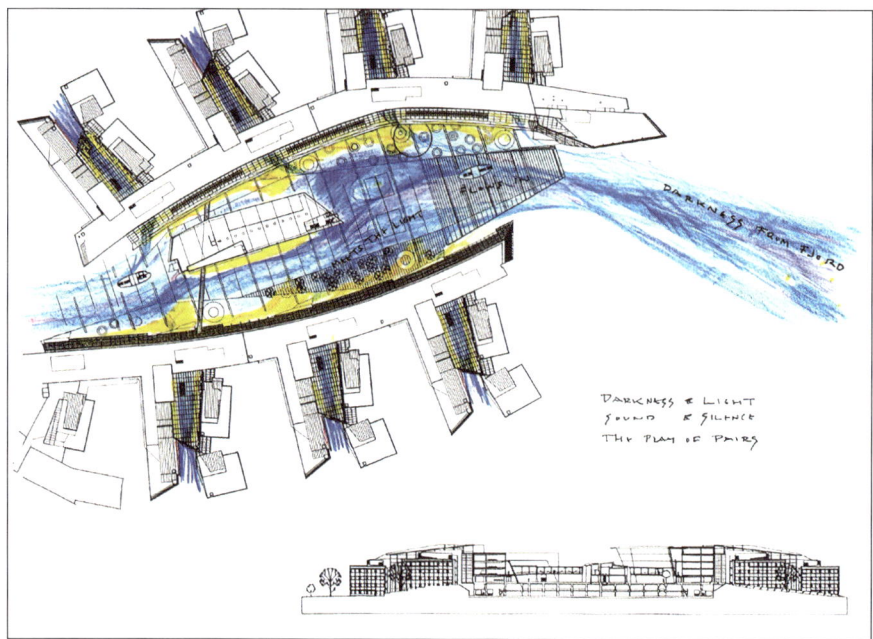

Figure 195: Telenor lighting design concept. The darkness of the Norwegian fjord was meant to flow silently through the building, back to freedom. Photo: Vesa Honkonen & Julle Oksanen

Love & hate; sound & silence; black & white; light & dark: These are some of the basic dichotomies of life. Without these opposing pairs, life would be terribly dull. "Shadow is light's best friend"—we desperately wanted to use this philosophy on that plaza.

The idea of bringing darkness from the fjord into the building and letting it flow through and back into the night somehow seemed very Norwegian. This concept immediately led us to conclude we did not want to install any lighting fixtures on the plaza itself. The huge glass openings in the building blocks would be our illuminators. The education center, which is located on the plaza, also had glowing walls.

This robust concept meant we would have to be able to influence and deploy office lighting to create this illumination.

Darkness Design master plan

The lighting design master plan occupies a strategic role in the overall design process. At this stage, lighting designers must find appropriate lighting hardware and software to implement the concept design on a practical level. If the right solutions cannot be found, it may be necessary to alter the concept or even create a new one.

Real-world experience brings maturity to a lighting designer's perception of what can be implemented and what cannot. Even if the concept proves difficult to implement as originally envisioned, one shouldn't give up hope. A small change in the concept may save a high-quality result.

The master plan phase consists of lighting fixture selection, lamp selection, preliminary checks on construction solutions, lighting calculations, computer images, etc. It excludes detailed design elements such as working drawings, construction detailing and lighting fixture integration into structural elements. This a good time to clarify exactly what is in the offer, to avoid lack of clarity later.

Figure 196: Telenor Lighting Design Master Plan, general computer image of the site. Photo: Telenor

To give a taste of what the design process looks like, we'll use the master plan design for the Telenor Headquarters in Oslo as an example. The master plan provides a deeper understanding of the building and its purpose, which guides the lighting design process in the right direction.

In this project, it was important to know:
- How people use the building during workdays (approaching the building, moving inside it, floors, cafés, offices, other activities, etc.).
- How the building is used at night.
- Materials of the building.
- Special needs of various activities.
- Who co-operates in the building, and their needs.
- National lighting design recommendations, how they are applied, and code requirements.
- How and to whom master plan designs will be introduced.

Figure 197: Telenor Lighting Design Master Plan. Computer analysis for the Atrium area. Photo: Erkki Rousku

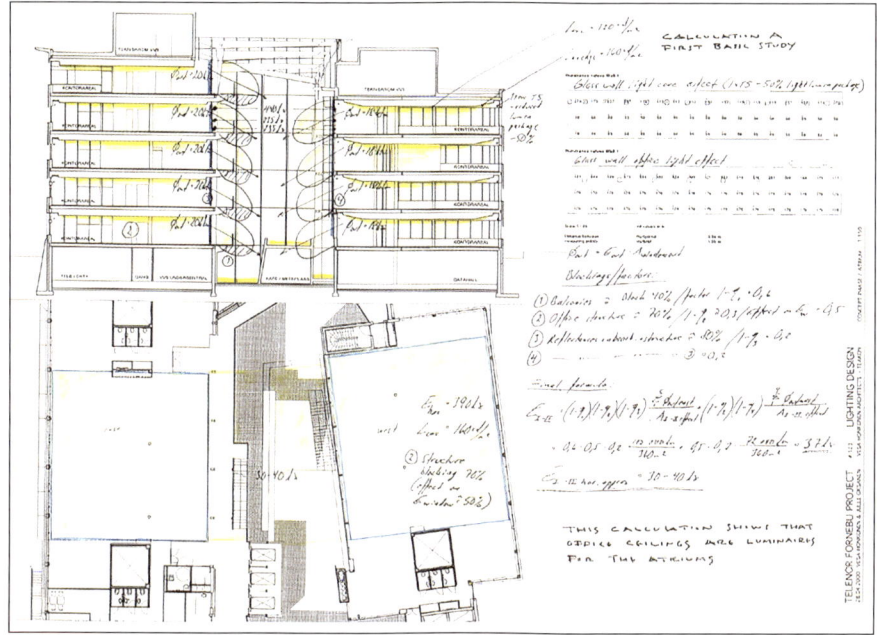

Figure 198: Telenor Lighting Design Master Plan. One result of our seven heuristic calculations. Photo: Vesa Honkonen & Julle Oksanen

The master plan phase starts with analysis. In existing buildings, one must analyze what's there now to be able to compare it to what's proposed. It's great to introduce the customer to the advantages of a new lighting design by comparing it to the old one.

New buildings call for more creativity in three-dimensional thinking. Designers must be able to imagine themselves in the space to develop fresh ideas and proposals. In good lighting design, three-dimensional thinking is essential to a successful result.

In the Telenor Building, we closed our eyes and thought about what it would be like to stand in the middle of the atrium and look around and up into the dark Norwegian night sky. Computer analysis finds that glass atriums often are "light traps" at night, with reflections bouncing around from one surface to another. This nudged us toward creating a lighting solution that minimized the mirror effects of light distribution surfaces, or luminaires, from hundreds of glass elements. The various angles of the huge glass elements, both horizontally and vertically, caused more problems. We also analyzed whether it would be possible to see out of the building at night.

We made a lot of heuristic calculations to solve all the above-mentioned problems. The reader is not expected to search for any details in those calculations presented here.

Office blocks are located on both sides of the glass atrium. We analyzed how much light actually flows from all those office areas to our design area—the glass atrium.

We asked the electrical designer who, together with luminaire manufacturers, was in charge of lighting in the office areas about the lighting procedure in those areas. Based on that information, we made lighting calculations for the offices to find the vertical illuminance values on windows located adjacent to the glass atrium.

We used the basic formula: $E = \Phi / A$. In words: The Illuminance value on window areas E is equal to lumen package of window areas Φ divided by illuminated window areas A. From this formula we solved the value of the lumen package Φ on window areas.

Then again, the basic formula $E = \Phi / A$ was used in a creative way: Now E was the illuminance value on the floor surface of the glass atrium, Φ was the lumen package that runs from the windows to the floor of the atrium, and A is the atrium floor area. Of course, we had to use additional factors (marked k in the formula), which affected the result. K is the product of the penetration factor of window glass, construction elements (balconies), direction of outflowing light from windows, reflections from surfaces and the approximated and "corrected" room factor of the atrium.

With these master plan calculations completed, we told the customer that these office designers had completed our work. We only needed to send the bill. We had made our calculations conservatively, to be on the safe side, and the result was 40 lx (4 fc). We told them that, in reality, it would be closer to 100 lx (10 fc) than 50 lx (5 fc). Of course, we wanted to create contrasts and functional lighting, not merely a "dull" 40 lx. We therefore said we had to design for more light but would take these values into consideration, too.

To avoid reflections of the light distribution surfaces of the luminaires, we had to create a new way to illuminate the glass atrium functionally. We could not use downward light luminaires, because they would have caused oval reflections from various angles and levels. We could not accept luminaires on the wall surfaces (again, huge reflections), nor wall washers (because of glass walls and difficult tracking systems).

"Normal" fluorescent luminaires were, of course, out of the question for the reasons mentioned earlier.

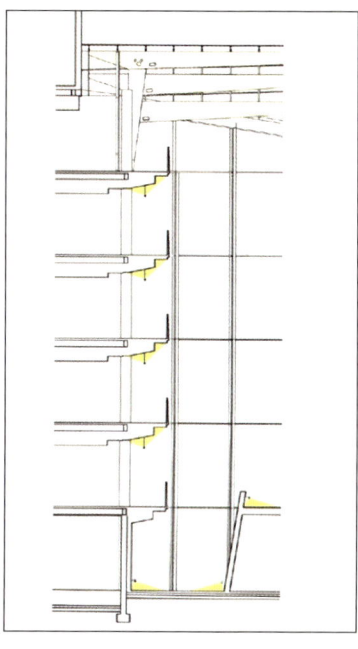

Figure 199: Telenor Lighting Design Master Plan. Atrium section. Indirect lighting solution. Photo Vesa Honkonen & Julle Oksanen

So, we created a new kind of functional lighting solution. We designed long rows of indirect fluorescent lighting, creating continuous light that functionally shone where it was needed while minimizing reflections.

We could not find available fluorescent luminaires providing continuous indirect light with the style, dimensions and invisible pendant and hanging systems we wanted, so we designed them ourselves. The Telenor building required 8 km (26,000 ft) of luminaires in rows, both indirect and direct versions.

Later on, this lighting system was to enter standard production for the manufacturer.

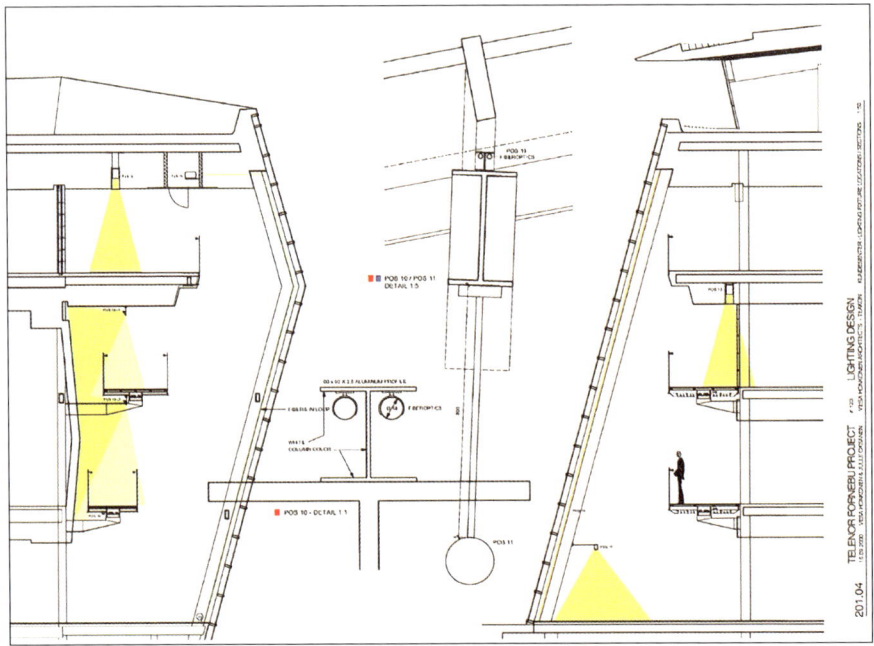

Figure 200: Telenor Lighting Design Master Plan. Boulevard sections. Hidden light distribution surfaces. Photo: Vesa Honkonen & Julle Oksanen.

When a project is large and complicated, all kinds of innovative hardware solutions may be needed to fulfill the concept. We also had some other solutions in addition to the row of fluorescent lights. Just a few examples of the solutions we used:
- "Black hole" luminaire.
- "Hidden lateral fiber optic" for the façade.
- "Fiber optics pylon".
- "Wall washing".
- "Deep down light" for orientation.

Remember that this was only the master plan at this stage, without details. In real-life projects, it's only necessary in this phase to show the client innovative solutions and help them follow the concept. Representatives of the client, like the architect, interior architect, electricity designer, etc., want to see that the project is feasible. Other designers need to be kept in the loop, because one designer's solutions can have a huge influence on other solutions. Co-operation is the key.

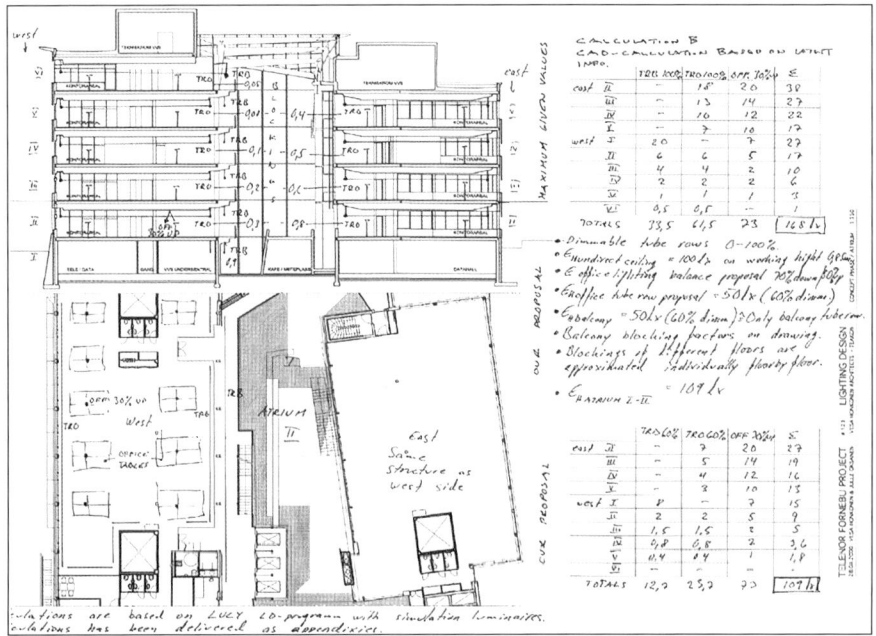

Figure 201: Telenor Lighting Design Master Plan. Dimming calculations. Photo: Vesa Honkonen & Julle Oksanen.

After all approvals on the mature master plan had been received from the client's side, we had to begin doing real-life calculations and deciding how to use light as a whole. The final calculations covered all related areas: offices, atrium floors, the café area, walking areas, areas in front of elevations, entrances, stairs and balconies.

As a reader, don't expect to glean any helpful details from these calculation figures. The main reason for introducing the calculations in this master plan context is to demonstrate how heuristic calculations supported our design work on this project.

We used the simple and free "DiaLux" lighting calculation program. It didn't include light distribution values for our new luminaire, so we had to execute simulation calculations by choosing a standard luminaire from the program. After that selection, we manipulated light distribution values to model our new luminaire. Based on these results, we came up with our proposal for how to use all lighting. We also had to take into consideration that these values would influence lighting values on the plaza, as well as the atrium areas.

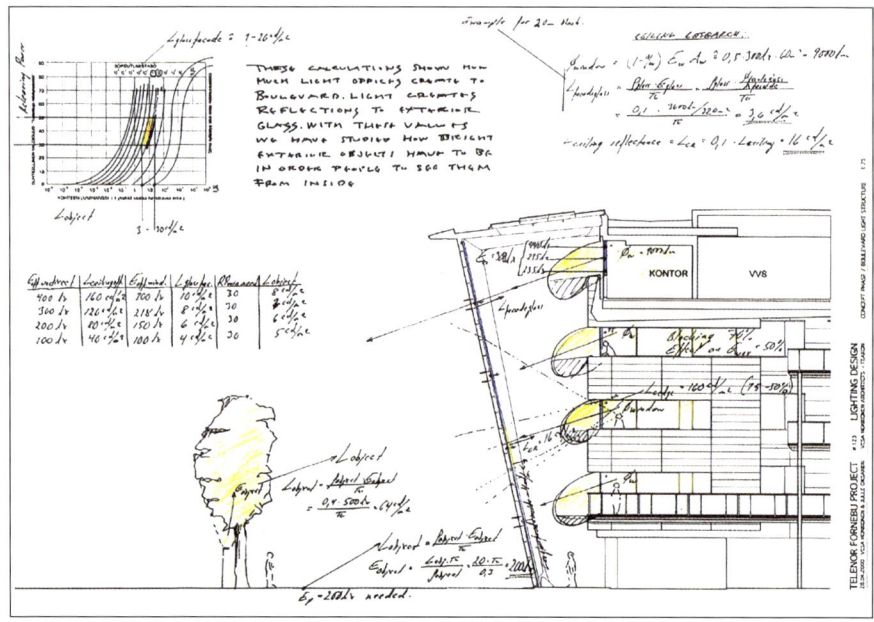

Figure 202: Telenor Lighting Design Master Plan. Contrast calculations to ascertain if people inside the building can see out during dark times. Photo: Vesa Honkonen & Julle Oksanen

The director of Telenor asked us whether his employees would be able to see out of the building during dark periods. That was a huge challenge for us as designers. We had to use Hopkinson's diagram as follows:

Mr. Svensson is looking out of his office window. He tries to look outside and enjoy the view.

There are two windows = mirrors between him and the plaza.

Let's assume first that Mr. Svensson has only outer glass elements (façade glass) and his own window is just a hole (equivalent to his office window being open).

What he sees is ambient light in the contrast ladder diagram (bunch of curves on left upper corner on figure). The value of that ambient luminance is reflections of interior values from façade glass. Those values are 1 – 20 cd/m2.

In order to "earn" 40 viewing points from left side of the contrast diagram, an object must have a luminance value of approx. 20 cd/m2. (Horizontal values on diagram). That means an E-value of approximately 200 lx (20 fc) on objects, for example the tree in the figure. This is not possible within reasonable limits.

Summing up: Mr. Svenson will be able to see out if Telenor purchases special glass that has a reflection factor of 2% (a very expensive glass material, normally used in control rooms). Otherwise, Mr. Svenson will only see a lot of reflections of his own face, wall bricks, structural elements and something outside. This would be a big mess.

Telenor decided to use normal float.

Figure 203: Telenor Lighting Design Master Plan. Computer image for plaza. Photo: Oliver Walter

This image was made for the client's directors and decision-makers, people who normally don't understand lighting "language." How you introduce project concepts and designs for these people can be a very sensitive issue. It is vital not to suggest that they are "dummies" when it comes to lighting, but rather to assure them that they are in charge. Visual images facilitate communication by helping them understand the goals.

The plaza luminaires are effectively the huge glass sides of both buildings. The glowing luminaires, hundreds of meters long and tens of meters high, give a soft general light to the plaza, with beautiful vertical values.

The eight entrances have their own inviting lights. The entrance canopies were illuminated by buried lighting units. The Education Center in the middle part of the plaza also creates a huge glowing luminaire, with light bridges to the buildings. Buried lighting units with asymmetrical light distribution were used.

The end of the building nearest to the fjord forms the so-called "PR part." It has blueish glowing façade lighting done with lateral fiber optics.

The parking area is located under the plaza. The two ovals are approach areas from the park to the plaza and the buildings themselves. Seven small dots on the left side are indirect luminaires leading from the outdoor parking area to buildings.

The greenish, round-shaped elements on the plaza are illuminated plant areas. They look small on this image, but each is many meters in diameter.

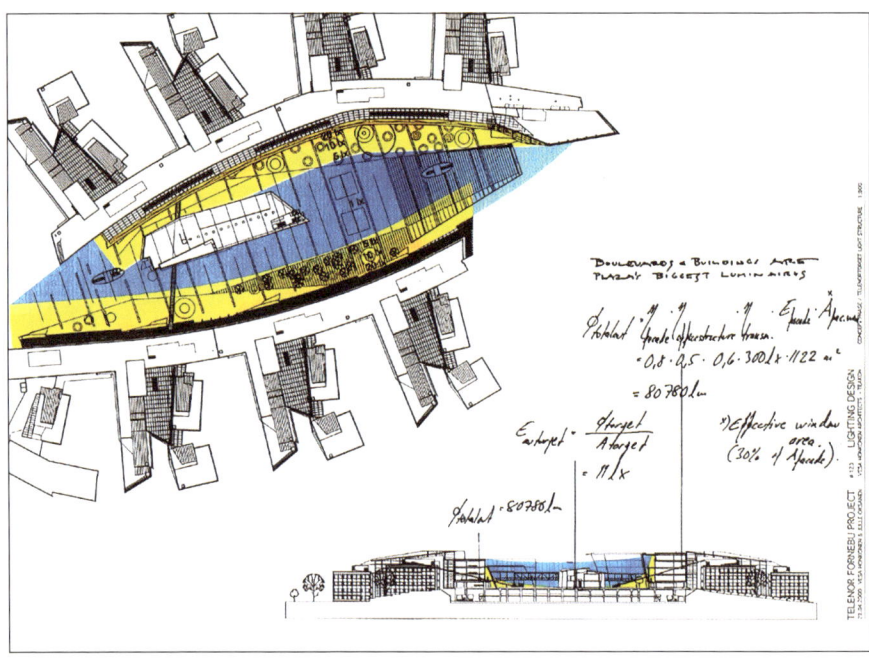

Figure 204: Telenor Lighting Design Master Plan. Heuristic plaza calculations. Photo: Vesa Honkonen & Julle Oksanen

The master plan calculation for the plaza was very interesting. Again, we used the basic applied lumen method to calculate illuminance values on the plaza. (The calculations shown in this master plan context are included merely to demonstrate how heuristic calculations supported our design work.)

In words, the formula is as follows:

The lumen package on the glass surfaces of both building blocks was calculated according to earlier calculations for offices.

This value was divided by A, the area of the whole plaza. ($\mathbf{E} = \Phi / \mathbf{A}$).

This value was multiplied by the sum of: kglass = glass penetration factor, kstructure = structural elements that reject light coming out of the building to the plaza (such as steel pylons, framing elements of glass, etc.), kdirectional = factor for how much of the light approaches the surface of the plaza.

$E_{averageplaza}$ = 11 lx (1.1 fc). This calculation indicated that light mostly fell down on the plaza near the facades, of course, fading beautifully near the center line. Vertical light is beautiful, because it doesn't disappear but flows vertically to the other side, flowing partly into the other building through the window areas, and partly reflecting back to the plaza and the other building again.

Result:

By minimizing reflections from windows, the design allows people in the atrium to see stars and other elements in the Northern sky. We were happy to see our concept come to life.

Indirect luminaire rows above the balconies offer a nice contrast and soft horizontal light on the floor level.

A nice contrast for café visitors is provided by a row of lights located very low on the left side of the image. The height of this row, placed on the small edge element of the café, is approximately two feet. The light distribution surface is towards the floor, but if needed it can be redirected.

Somehow a cozy feeling is created in the office environment.

Figure 205: The result of the Telenor lighting design. The café in the atrium.
Photo: Jan Drablos

Figure 206: The result of the Telenor lighting design. Stairs. Photo: Jan Drablos

Rows of lighting were installed approximately one foot above the stair surface. The lines follow the angle of the stairs, and the light distribution surface is aimed towards the stair treads.

This solution for the stairs looks and sounds easy, now that the project is completed, but it was problematic in the real design phase. One reason was that nobody had ever used this method to illuminate stairs.

The importance of continuous light on the balcony can also be felt on the right side of this image.

Rows of "light tubes" provide anonymous and functional elements in the space. They do not clash with any architecture but still fulfill their task as luminaires.

Contrast and a continuous row of light "tubes" somehow lend a certain charisma to the space.

Figure 207: The result of Telenor lighting design. Atrium. Photo: Jan Drablos

Nice illuminance values, which fulfill lighting recommendations in Norway while offering great contrasts, make for an interesting combination. Shadow is light's best friend.

Figure 208: The result of the Telenor lighting design. Plaza. Photo: Jan Drablos

Seen from the other direction, the plaza is cozy despite being huge, an unusual combination.

The vertical illuminance values create a beautiful ambience and make it possible to recognize approaching people's faces and expressions at a reasonable distance. This is an important factor in dark environments. According to research by the anthropologist E.T. Hall and engineer Wout van Bommel, to achieve this ease of recognition, the semicylindrical illuminance value at a height of 1.5m (5 ft) must be E_{scmin} = 0.8 lx (0.08 fc). At this value, an observer can recognize the faces of approaching people from the mandatory recognition distance of 3.5m (11.5 ft). The lighting values are modest, but being surrounded by illuminated walls imparts a feeling of safety. The contrast of people seen moving around, even at a distance, creates a sense of comfort.

Lighting design details

Such big projects need a professional coordinator, especially when the construction process is moving ahead quickly. In this case, the project coordinator was a talented and experienced project leader, Jan Drablos of Multiconsult Norway. He transferred the Lighting Concept and Master Plan information to the electrical engineering consultants whose task was to add lighting solutions as part of electrical drawings prepared for the electrical contractor. In the detailed design phase, the lighting designer has only a small role, especially in this kind of big project. In the implementation phase of this project, Jan Drablos took full control of the whole lighting design and procurement, even though it was a technical turnkey contract. This was to avoid being cheated on the types of fixtures to be installed (a common issue all over the world). Multiconsult decided on the manufacturers and types of all fixtures. Testing was often carried out before a decision was made on the type and installation. The electrical engineer consultant team was instructed to perform the final detailed design according to the lighting designer's concept and Multiconsult's choice of lighting units, with great success.

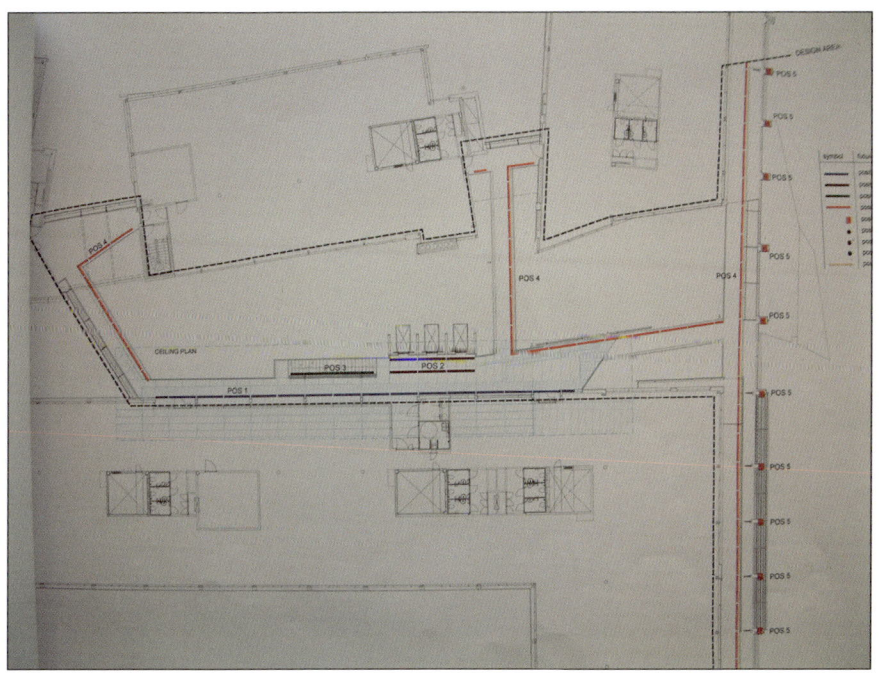

Figure 209: Detailed design drawing of Telenor atrium. Lighting positions. Photo: Jan Drablos, Multiconsult.

Our lighting project was completed successfully with only one luminaire type designed by ourselves (Vesa Honkonen & Julle Oksanen). Notor was manufactured by international luminaire manufacturer Fagerhult Belysning AB. Fagerhult is one of Europe's leading lighting companies, with 2,200 employees in 20 countries. Fagerhult develops, manufactures and markets innovative and energy-efficient lighting solutions for professional indoor, retail and outdoor environments.

Telenor has over 8000 meters of Notor luminaires.

DEMONSTRATIONS AND IMPLEMENTATION

Demonstrations are important before final implementation. Computer images or calculations do not substitute for seeing the 3-dimensional effect of designed solutions in real life.

Figure 210: Façade lighting demonstration. Photo: Jan Drablos, Multiconsult

Final thoughts on Telenor

This Darkness Design project is a great example of how thinking in terms of a heuristic metaphor leads the process in a unique direction.

The created concept falls somewhere between the analytical approach and the Gestalt approach.

As a result of this concept, Fagerhult AB manufactured and brought to market the first linear light product, Notor. Some 11 km of Notor light lines illuminate the whole building and its plaza. This was a direct outcome of a new "whispering light" metaphor, part of modern Darkness Design.

APPENDIX 2: FUTURISTIC "CITY 2030"

Figure 211: Hypothetical large-surface lighting in a city in 2030: Philosophy and lighting design by Julle Oksanen. Photo: Oliver Walter and Dan Silberman

Let's imagine the lighting in a dream city professionally designed by architects and engineers well trained in lighting: "City 2030."

In 2030, all schools of engineering and architecture have architectural Darkness Design education as a natural and compulsory component of their programs. By 2030, the focus will have become *quality* lighting design instead of old-fashioned *quantity* lighting design. Solar-powered lighting units will be in widespread use, supplying free energy for road lighting networks using individual inverters active during daytime. Large, fully dimmable LED panels, covered with abrasive-blasted double glass with light-scattering film between, will provide soft and efficient light. Road lighting will have become history, and dynamic vehicle traffic lighting systems and "glow-in-the-dark" road markings will be in common use in all over the world. Architects will use various kinds of lighting design processes and metaphors daily in their work, leaving the technical details to engineers.

Ambient lighting in the City 2030 structure

Glaring road lighting and ugly lighting fixtures on poles have been converted to modestly glowing large-surface light-emitting planes located in city structures, for example integrated into buildings. Glowing panels are a natural part of city structure and beautification. Light is produced from emitting distribution surfaces equipped with less than a 300 cd/m2 "brightness package." Brightness is 1/100 compared to old-fashioned road lighting units. Even the surface of the moon looks brighter than City 2030 road lighting solutions. (Measured moon luminance is 2000–2500 cd/m^2, according to Oksanen–Gabriel measurements in Canada and Finland in the 1990s.)

All large-surface light-emitting elements have LEDs as light sources, while the upper part of the structure is fully covered with solar panels. They produce free energy for power plants through their inverters. (Marked luminaires 1-11 in the graphic appearing ahead in the section on lighting calculations could annually produce 428,000 kWh free solar energy if the project was in southern Finland, where solar energy availability is 1000 kWh/a). When night falls and road lighting systems turn on, the inverters are switched off automatically from the power plant's remote control rooms. In the City 2030 vision, the lighting system produces more energy than it consumes.

All lighting elements are equipped with intelligent LED drivers. When there is no movement in a space, the lighting level can be adjusted automatically to a pre-set lower luminance value, for example 10%. When motion detectors recognize something moving in the space, the light will automatically increase to a higher pre-set lighting value (for example, 80% of maximum). Graduated shadow design can easily be done using pre-set LED fixtures after installation has been accomplished. Shadow zoning can be created, for example, block by block.

'Focal glow' and 'play of brilliants' in City 2030

In City 2030, Darkness Design teams have been educated not only to create beautiful shadow gradation zonings in "ambient light" areas, but also to balance the space with the right kinds of "focal glow" and "play of brilliants" elements. Shop windows and illuminated advertisements will be professionally done. These harmonized elements create a city center for City 2030 that will feel like a living room, with professionally designed beautification.

Figure 212: Modern architecture supports large-surface lighting solutions. Photo right: Hypothetical Dream City 2030. Photo: Julle Oksanen. Photo left: Leppävaara business area, Espoo, Finland. Photo: Julle Oksanen. Oksanen

Structure of "City 2030"

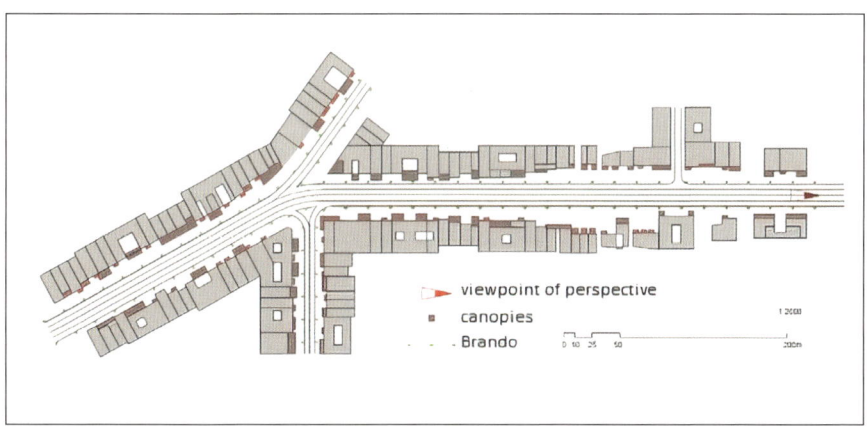

Figure 213: City 2030 Plan drawing: Design Julle Oksanen. Photo: Oliver Walter

Lighting calculations
Calculation area

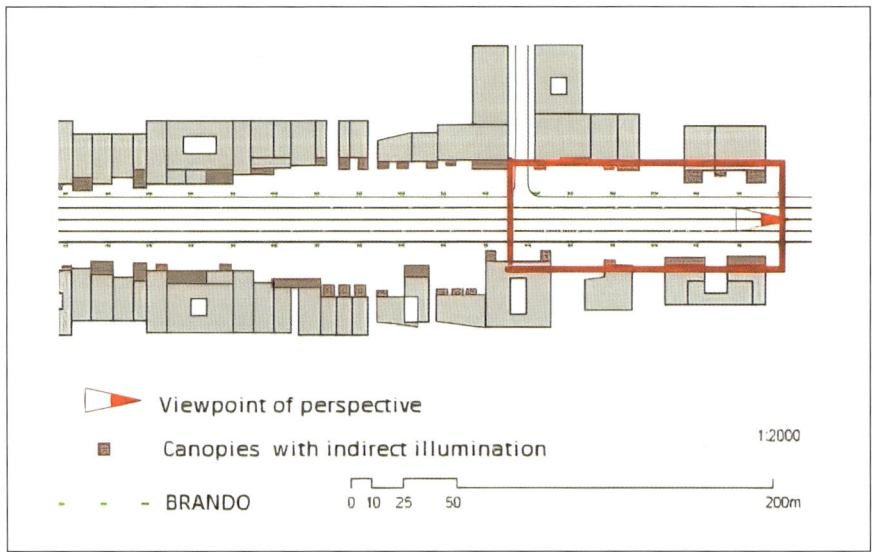

Figure 214: Selected lighting calculation area. Photo: Oliver Walter.

Luminaires used in calculation

1. System FLAT LED

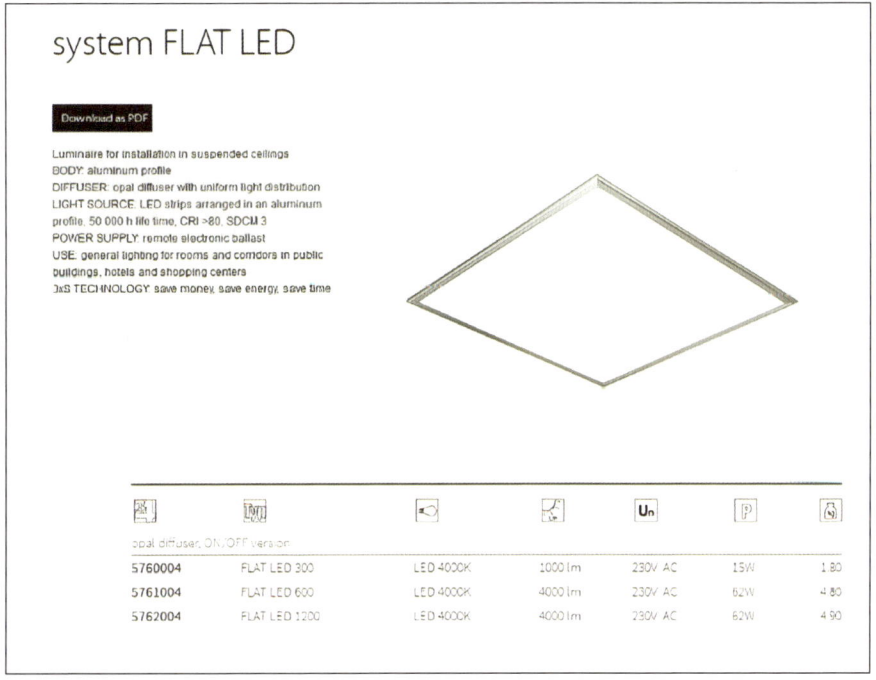

Figure 215: Simulation luminaire for large-surface lighting elements. Photo: ES-System.

The large-surface luminaire selected is an ES-System "system FLAT LED" luminaire. In the calculation, the light-emitting area has been enlarged to obtain reliable calculation results. The lumen output value in each light element (luminaires marked 1-11 in the graphic appearing below) has been calculated according to the size of the light-emitting surface and the luminance value of each light element (1000 cd/m^2). The light distribution of the simulation luminaire is equal to the total light distribution of all the light elements.

2. Brando luminaire

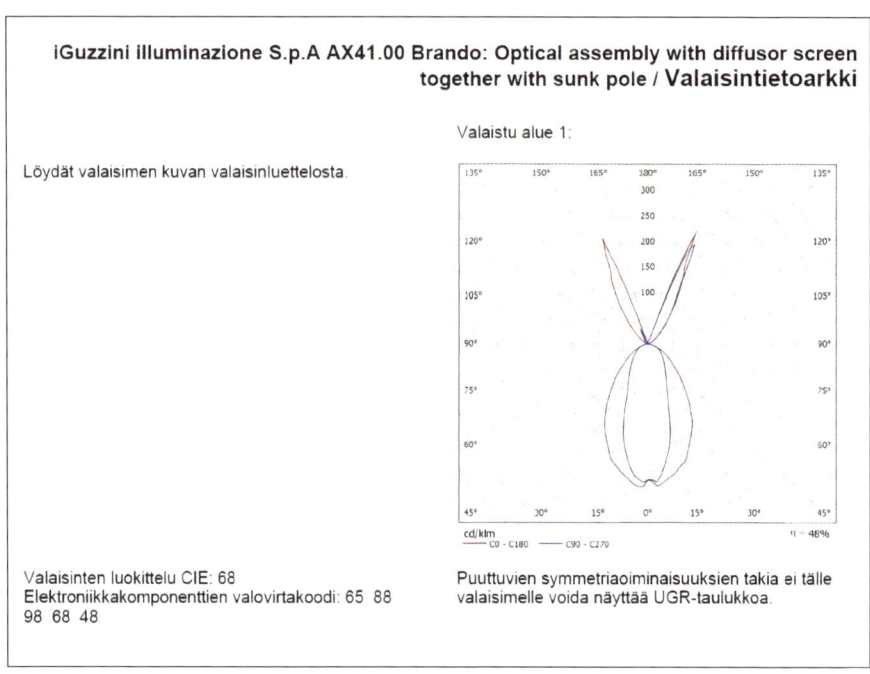

Figure 216: Light distribution for Brando luminaire. Photo: Iguzzini Brando brochure.

Luminaires and basic calculation values

Large-surface luminaires selected for the calculation are marked in next figure 148 (Luminaires 1–11 + Brando luminaires in other calculations). The basic luminance value selected for all luminaires in all calculations is 1000 cd/m². The final "brightness" values of the light-emitting surfaces can be determined by multiplying the results of the calculation by the selected luminance value divided by 1000 cd/m². For example: 1000 cd/m² light-surface luminance gives 60 lx semi-cylindrical illuminance (Esc). We want average Esc = 6lx on the whole area. This means that we need 100 cd/m² as the final luminance for the emitting surfaces (a measure of how bright the surfaces look).

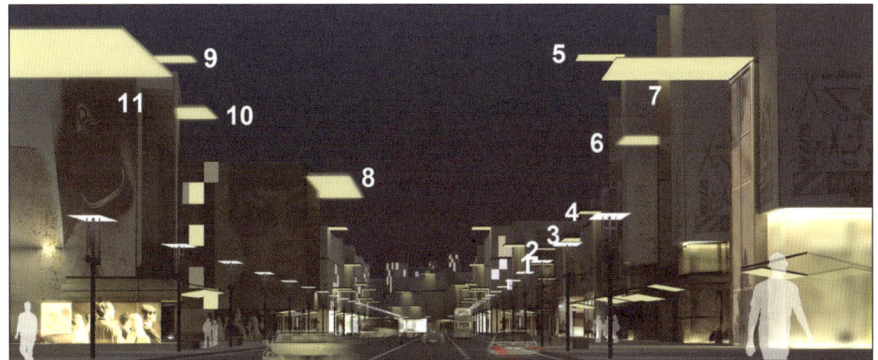

Figure 217: Luminaire numbering in calculations: Design Julle Oksanen. Photo: Oliver Walter and Dan Silberman.

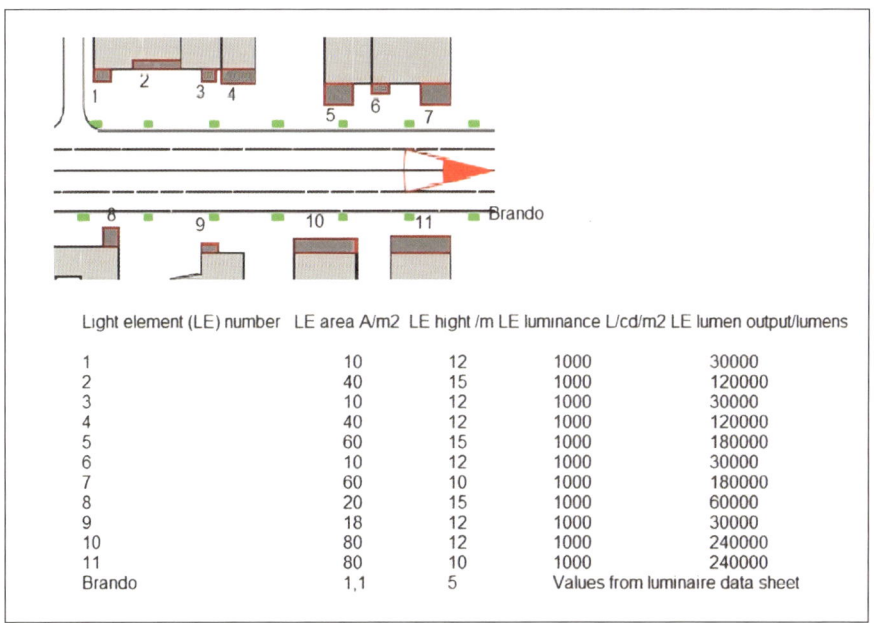

Light element (LE) number	LE area A/m2	LE hight /m	LE luminance L/cd/m2	LE lumen output/lumens
1	10	12	1000	30000
2	40	15	1000	120000
3	10	12	1000	30000
4	40	12	1000	120000
5	60	15	1000	180000
6	10	12	1000	30000
7	60	10	1000	180000
8	20	15	1000	60000
9	18	12	1000	30000
10	80	12	1000	240000
11	80	10	1000	240000
Brando	1,1	5	Values from luminaire data sheet	

Figure 218: Large-surface lighting elements and basic calculation values. Only the luminaires mentioned (1-11 + Brando) are included in the calculation. Large-surface emitting planes on the lower level and the light from windows are excluded from the calculations.

Calculations for basic values without Brando luminaires

Figure 219: City 2030 without Brando luminaires: Design Julle Oksanen. Photo: Oliver Walter and Daniel Silberman

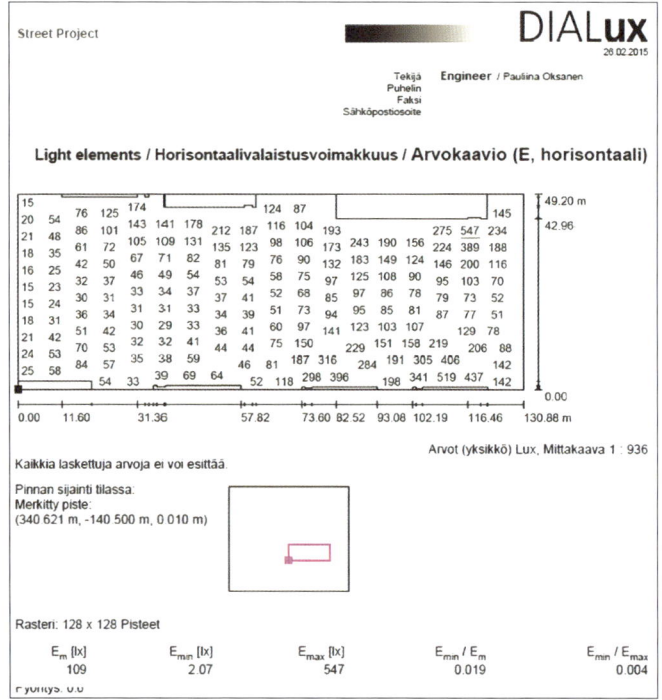

Figure 220: Horizontal illumination values on the surface level, road and pedestrian. The calculation is done using the DiaLux Street Lighting design calculation program. Lighting calculation by engineer Pauliina Oksanen.

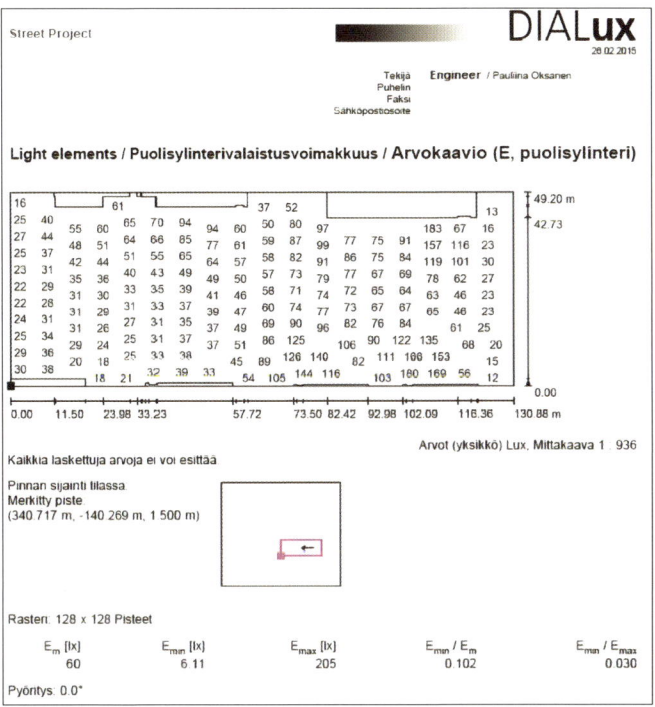

Figure 221: Semi-cylindrical illuminance values (Esc) at average face height of 1.5 m. The calculation is done using the DiaLux Street Lighting design calculation program. Lighting calculation by engineer Pauliina Oksanen.

Calculations of basic values with Brando luminaires

Using Brando luminaires is an option and has little influence on the calculation values because large-surface structural lighting emitters are so big, and they have a lot of lumens (how many liters of light emitters produce). The Brando luminaires in this specific example offer a sense of rhythm and human scale.

Figure 222: City 2030 with Brando luminaires: Design Julle Oksanen. Photo: Oliver Walter and Dan Silberman

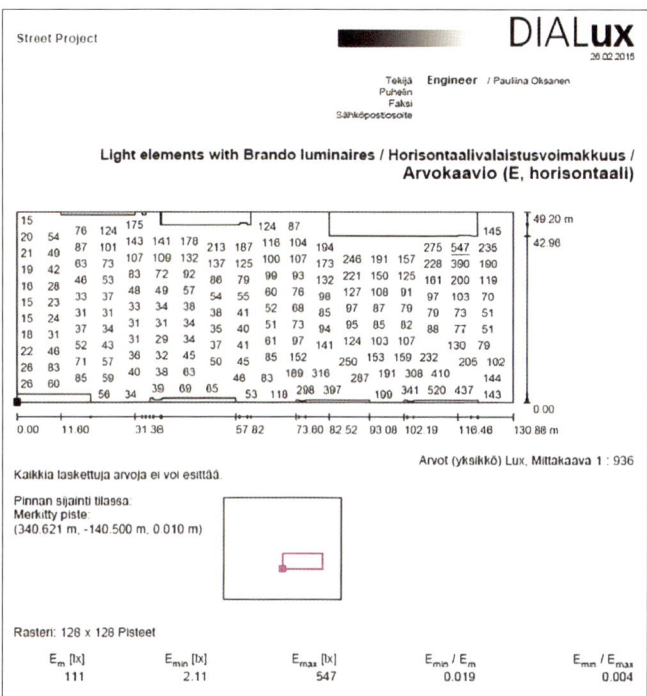

Figure 223: Horizontal illumination values at the road and pedestrian surface level. The calculation is done using the DiaLux Street Lighting design calculation program. Lighting calculation by engineer Pauliina Oksanen.

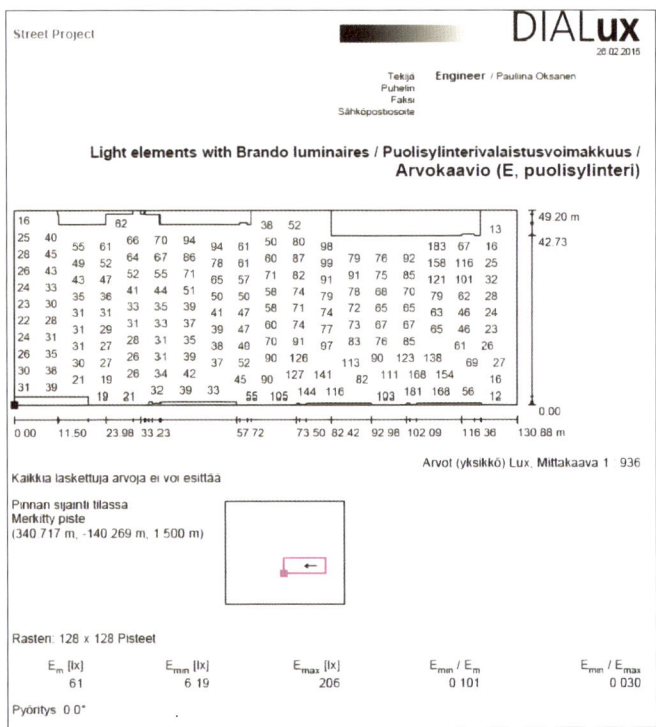

Figure 224: Semi-cylindrical illuminance values at an average face height of 1.5m. Calculation is done by using the DiaLux Street Lighting design calculation program. Lighting calculation by engineer Pauliina Oksanen.

Creating the public "living room"

The City 2030 lighting solution offers a new metaphor for light that is fascinating and hopefully becomes part of modern city beautification in the future. This is a great biomimetic application, where the brightness of the light-distribution surfaces of the luminaires imitates the brightness of the moon and a cloudy sky.

As has been reiterated several times in this book, shadow is light's best friend, while glare is light's worst enemy. The vital reason to deploy large-surface emitting lighting solutions is that this offers a combination of total freedom from glare and full adjustability, allowing the designer to create gradations of shadow. Modern LED technology and intelligent control units allow this privilege in many kinds of practical lighting design solutions. Relaxed movement in nocturnal spaces is possible by using sufficient Esc- values and minimized LA ^{0,25} glare values.

The minimum Esc- value is 0.8 lx. This means that, with or without using Brando luminaires, 10% of the calculated result is more than enough. (Esc without Brando luminaires is 6.0 lx; with Brando luminaires 6.1 lx.). This allows designers to adjust shadow gradations so that an intersection, for example, receives more light than straight stretches of road (even double the amount of Esc means only 20% of calculated values), and 10% of the Esc- value of the calculations also means that large light-emitting surfaces only glow—and LEDs could last for more than 100 years without maintenance.

Diagnosing disturbing glare values with the formula LA 0,25

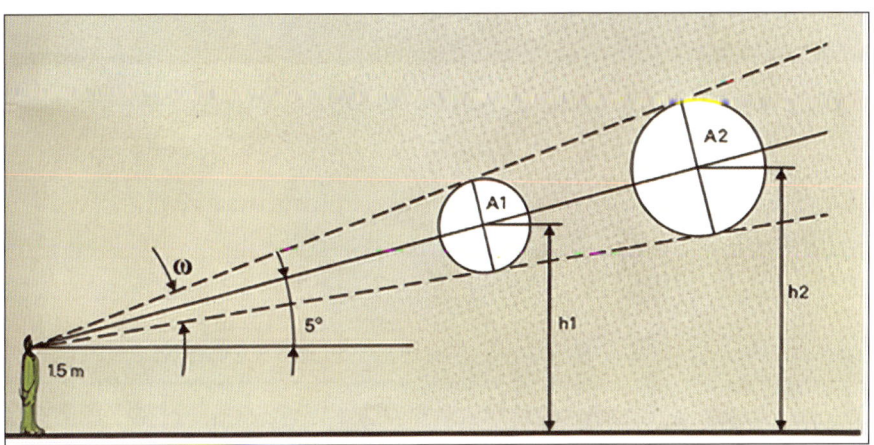

Figure 225: Principle drawing for LA 0.25- value. (Philips ILR 1980 /3)

LA 0.25
where,
L = measured luminance between angles of 85-90 degrees from vertical (light-emitting area of the luminaire light-distribution surface),
A = bright luminaire light distribution section.
The maximum values depend on the height of the luminaire:
LA $^{0.25}$ = 1250, h < 4.5m
LA ^{0,25} = 1500, 4.5m < h < 6m
LA ^{0,25} = 2000, h > 6m.

If light elements are higher than 6 m, this means that LA ^{0,25} –values must be less than 2000.

Calculations:

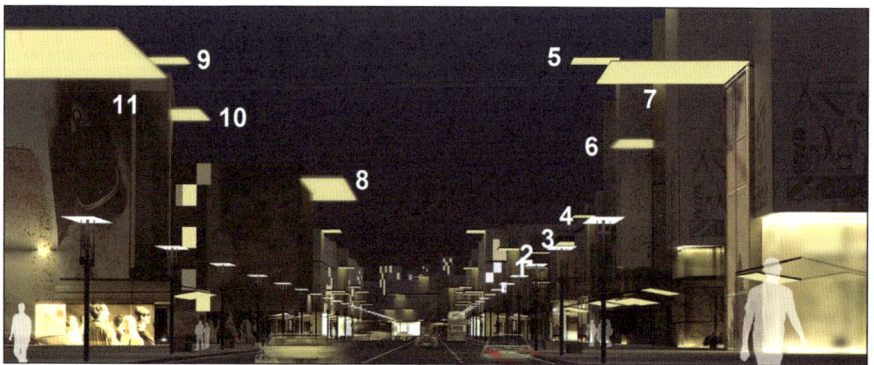

Figure 226: Luminaire numbering for LA^{0,25} calculations: Design: Julle Oksanen. Photo: Oliver Walter.

Luminaire	Ara A (m²)	Hight (m)	Luminance (cd/m²)	LA^{0,25}
1	10	12	100	97
2	40	15	100	137
3	10	12	100	97
4	40	12	100	137
5	60	15	100	151
6	10	12	100	97
7	60	10	100	151
8	20	15	100	115
9	10	12	100	97
10	80	10	100	162
11	80	10	100	162

Total combined area of the light-emitting surfaces: 420 m²
Sample calculus: Luminaire number 10:
Area seen from an angle 5 degrees from vertical:
A projection = A x sin 5

A projection = 80 m² x 0.087 = 6.96 m²
Luminance of emitting surface = 100 cd/m²
LA 0,25 = 100 x 6,96 0,25 = 162,4 <<< 2000 (which is already a great value)

Personal zoning and mandatory recognition distance
Esc value rationale

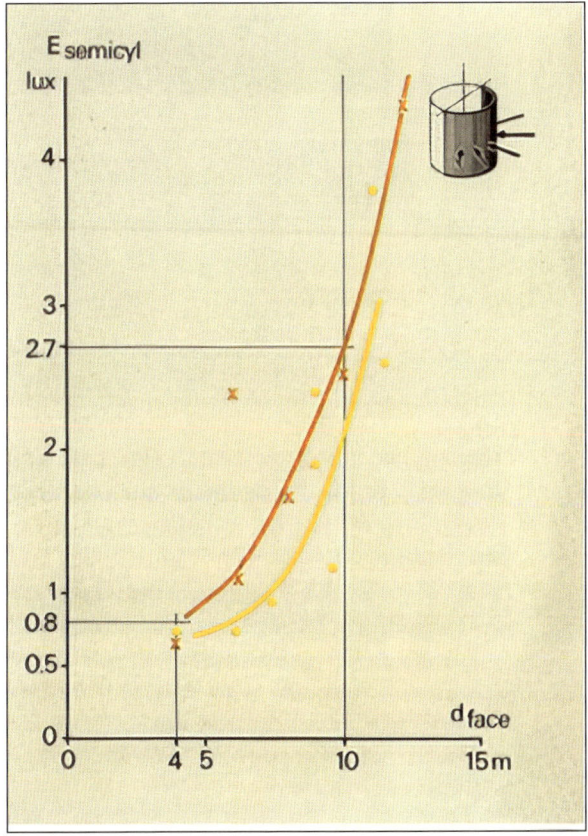

Figure 227: The minimum Esc- value for recognizing a face when approaching in darkness is 0.8 lx. (Philips ILR 1980/3). In our solution, the Esc- value is 6 lx. That means that the face-recognition distance is over 15 meters, guaranteeing relaxed movement for pedestrians.

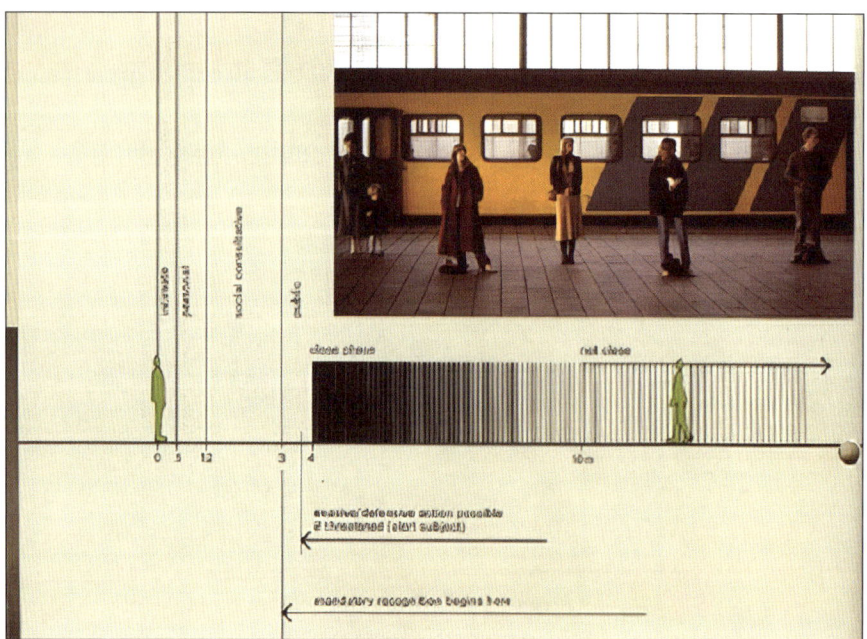

Figure 228: Personal zoning, according to research. (Philips ILR 1980 /3). In City 2030, recognition distance is over 15 m, while the mandatory recognition distance is only 3 m.

Heuristic energy-saving calculations using "rule of thumb"

1st Step:

Let's say our hypothetical City 2030 is located in Eugene, Oregon, USA. First, we need a figure for annual solar energy. According to the figure below, this is 6kWh/m²/Day/Year.

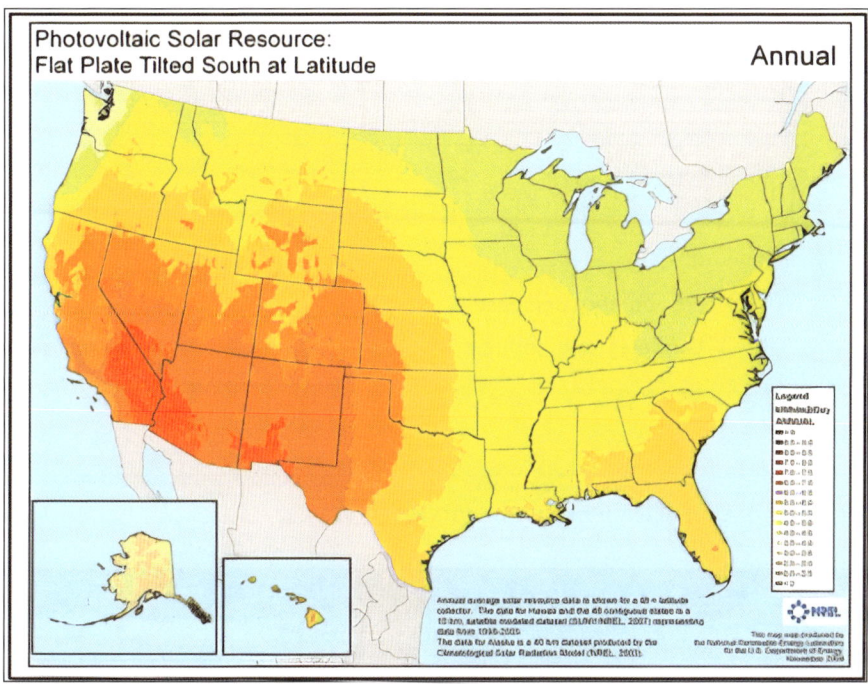

Figure 229: PV Solar Radiation (10 km) Static Maps (1998 to 2005 data). Photo: U.S. Department of Energy 2008.

2nd Step:

Calculate the total area of solar panels on the calculation area (area includes 11 large-surface luminaires). **It is A= 420 m2**

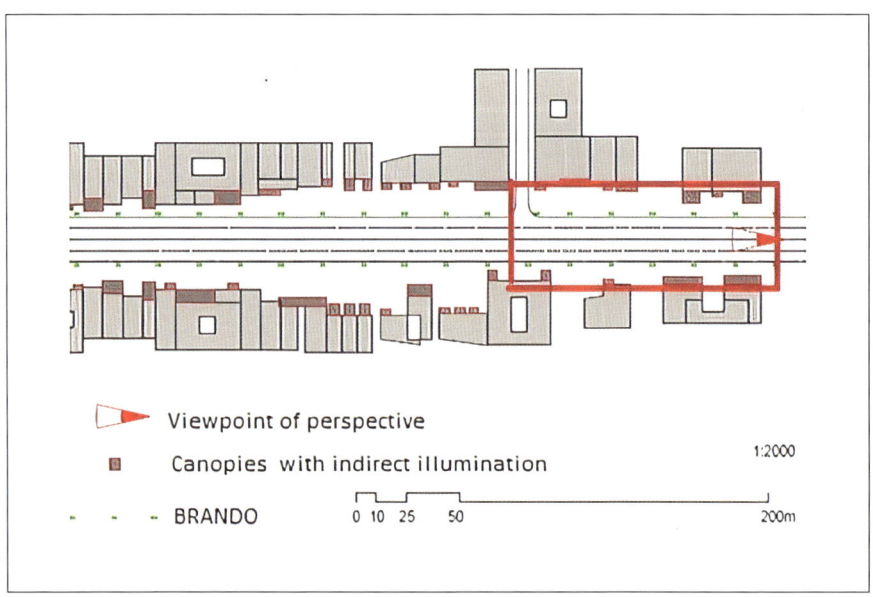

Figure 230: Calculation area. Photo: Oliver Walter

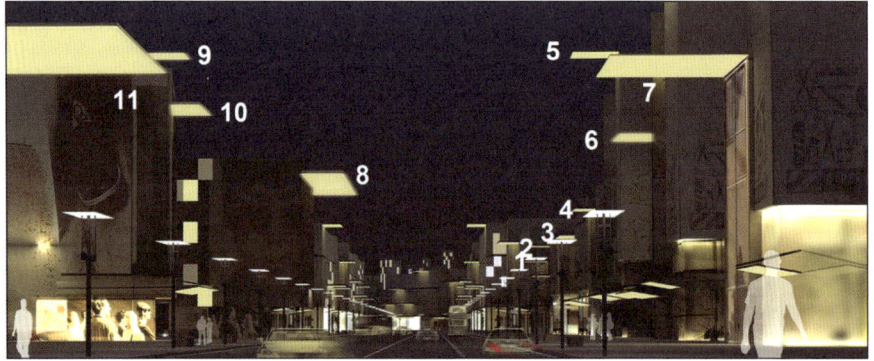

Figure 231: Large-surface luminaires numbered 1-11. Photo: Oliver Walter.

Luminaire	Area A (m²)	Height (m)	Luminance (cd/m²)	LA^0,25
1	10	12	100	97
2	40	15	100	137
3	10	12	100	97
4	40	12	100	137
5	60	15	100	151
6	10	12	100	97
7	60	10	100	151
8	20	15	100	115
9	10	12	100	97
10	80	10	100	162
11	80	10	100	162

Total area 420 m2

3rd Step:
Calculate the total annual solar energy/day garnered by these 1-11 large surface solar panels.

6kWh/Day/m2 x 420m² = **2520 kWh/Day.**

4th Step:
Calculate the total electrical power and energy for LED panels, which produce the needed lighting level in the calculation. The formula is:

$$L = \frac{T \times \Phi / A}{\varpi}$$

where:

L is the luminance of LED panel surfaces (100 cd/m²)

T is the penetration factor of the LED panel glass (0.5, 50% of lumen flux penetrates the glass)

Φ is the total lumen flux that flow on to the inner parts of all glass surfaces (must be calculated).

A is total area of LED-emitting surfaces (420m²)

-ϖ = 3.14

After simple calculations, the total lumen flux from all LED luminaires that flows onto the calculation area is: Φtotal = 263760 lm.

Because the LED efficiency is 50 lm/W, total electric power is 263,760 lm/50 lm/W = 5275 W

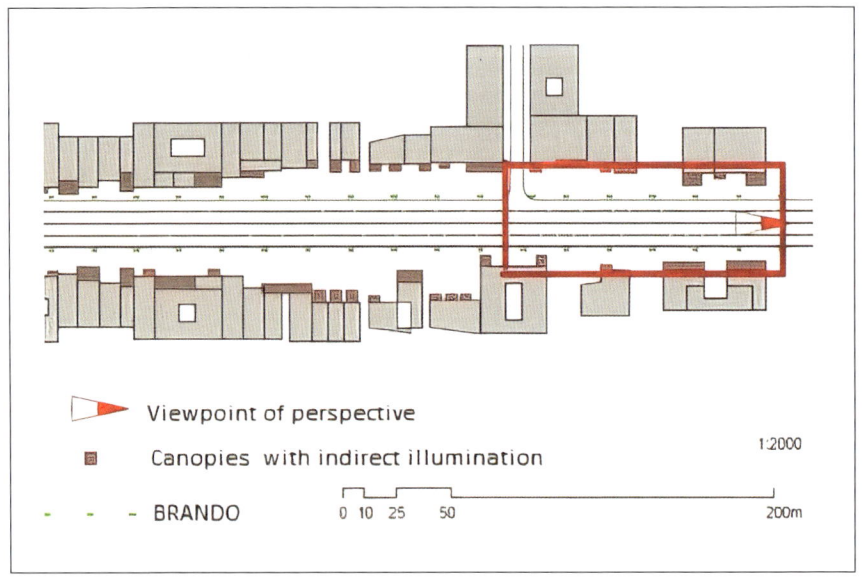

Figure 232: 5.3 kW is needed to power LED panels located in the calculation area (marked RED)

The lights are on 10 hours/day. So the total needed energy is **53 kWh/day/year.**

5th Step:

The energy for for lighting and how much is left over can now be calculated. The total solar energy was **2520 kWh/day** and the lighting needs **53 kWh/Day**. This means that 2467 kWh/day is free to be returned to the electricity grid from the area calculated.

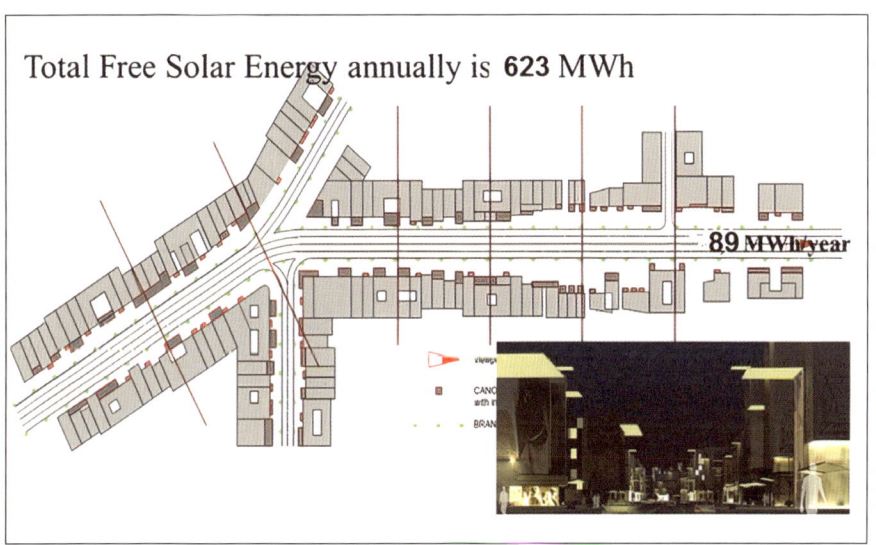

Figure 233: This project is not only a large-surface luminaire project, but also a daylight electricity generation project. (7 zones x 89MWh/zone = 623 MWh)

The system efficacy when supplying energy from a modern solar panel system to the electric grid is approximately 25%. This means that, heuristically calculated, this whole City 2030 project annually yields 155 MWh free solar energy, after part of the generated power has been used on the lighting system itself.

ABOUT THE AUTHOR

Julle Oksanen and architect Professor Dr. Hannu Tikka are part of Group X, a focus group at the Department of Architecture at Aalto University. They form a "Light & Space Academy," part of the Finnish Mobile University, arranging travelling educational activities in the USA, Asia and Europe. Julle Oksanen graduated with a PhD from the Department of Architecture, Aalto University. He researches the relationship between light pollution, Darkness Design, architecture and biodiversity.

Before his work in Darkness Design, he spent more than 30 years as a lighting and fixture designer, and his company, Julle Oksanen Lighting Design Ltd., has carried out diverse projects that included a cathedral, amphitheater, airport, city centers, office buildings, etc., in the United States, Europe and Asia. He has also designed luminaires for international manufacturers, both custom and for standard production. In addition to his PhD from Aalto University, he has a Master of Science degree in Landscape Architecture from the University of Tennessee College of Architecture and Design. He is also a Professional Engineer (electricity) as a graduate of Helsinki Technical College. He has held visiting professorships and other academic positions at universities in London, New York, Ithaca, Philadelphia, Oregon, Tennessee and Norway. He has lectured at important Lightfairs in New York and Las Vegas and in over 20 universities around the world.

His dissertation "Design Concepts in Architectural Outdoor Lighting Design, Based on Metaphors as a Heuristic Tool" has been honored as the best final thesis in the field.

RELATED TITLES BY HANCOCK HOUSE

Solid Air: Invisible Killer
Saving Billions of Birds from Windows

Klem, Daniel Jr.
978-0-88839-646-4 [Trade Paperback]
978-0-88839-640-2 [Trade Hardcover]
978-0-88839-665-5 [eBook]
6 x 9 sc, 224pp
$24.95

As useful and attractive as sheet glass and plastic are in human buildings, the windows they make are indiscriminately lethal and devastating to free-flying birds.

This book describes the cause and breadth of this universal problem and how to solve it. Detailed objective observations and experiments reveal that birds behave as if clear and reflective windows are invisible to them. Alarmingly, among the dead are the fittest individuals
in species populations.

Science has documented that a decline of a third of the annual North American bird population, approximately 3 billion individuals, has occurred since 1970, and one of the principal mortality sources is windows. However agreeable and possible, citizens the world over are asked to take action and join in making the human built environment safe for bird life as one of Nature's most beautiful, useful, and spiritually uplifting creations of planet Earth.

Unlike the complexities of other environmental challenges, such as climate change, this important conservation issue for birds and people can be solved, and the means to do so are described within the pages of this work to guide this worthy effort.

Hancock House Publishers
19313 Zero Ave, Surrey, BC V3Z 9R9
www.hancockhouse.com
info@hancockhouse.com
1-800-938-1114